Advances in Cycle Time Management

Tin-Chih Toly Chen

Advances in Cycle Time Management

Springer

Tin-Chih Toly Chen
Department of Industrial Engineering
and Management
National Yang Ming Chiao Tung University
Hsinchu, Taiwan

ISBN 978-3-032-06910-8 ISBN 978-3-032-06911-5 (eBook)
https://doi.org/10.1007/978-3-032-06911-5

This Springer imprint is published by the registered company Springer Nature Switzerland AG
The registered company address is: Gewerbestrasse 11, 6330 Cham, Switzerland

Competing Interests The author has no competing interests to declare that are relevant to the content of this manuscript.

Contents

Chapter 1
Cycle Time Management

Abstract This chapter first defines the basic concepts, such as cycle time, step cycle time, and remaining cycle time, and discusses the relationship between these times. Subsequently, cycle time management is defined and its six activities are introduced: cycle time prediction, cycle time modeling and analysis, management support, cycle time control, cycle time reduction, and cycle time learning. Relevant cases or references are reviewed to illustrate the importance and implementation of each activity. From the review results, the most studied cycle time management activities include cycle time reduction, cycle time modeling and analysis, and management support. In addition, difficulties faced by cycle time management are also summarized. Recently, some advanced information technologies have emerged that can be used to overcome the difficulties of cycle time management and improve its effectiveness: Industry 4.0, big data analytics, edge computing, cloud computing and ubiquitous computing, explainable artificial intelligence, etc. How these advanced information technologies can help improve the effectiveness of cycle time management are discussed. References supporting these potential applications are also reviewed.

Keywords Cycle time · Cycle time management · Big data analytics · Industry 4.0 · Edge computing · Explainable artificial intelligence

1.1 Cycle Time

Cycle time, **flow time**, or **manufacturing lead time** is the time required to complete all operations of a job in a factory, which may be delayed by factors (or events) such as machine downtime (planned or unplanned), product quality issues, ineffective job release or scheduling, operator inefficiency, rush orders (or jobs), and inconsistent human intervention [1]. A few references distinguish between cycle time and flow time by removing waiting time from cycle time so that cycle time equals total processing time [2]. Another related term is takt time, which is the rate at which a manufacturing system needs to work and output to meet customer demand without accumulating inventory [3]. Namely, takt time is the required inter-output time. In a

T.-C. T. Chen, *Advances in Cycle Time Management*,
https://doi.org/10.1007/978-3-032-06911-5_1

lean manufacturing system, takt time is the target for setting the step cycle time of each step [4].

For some products, such as semiconductors, the cycle time of a job can be very long, often up to several months [5]. On the contrary, large products may have short assembly time (i.e., cycle time), while their component suppliers have long manufacturing time. For example, the cycle time to assemble a Boeing 737 MAX is only 9 days [6], while preparing its parts may take several months.

The output time of a job from the factory can be derived by

$$O_j = R_j + C_j \tag{1.1}$$

where O_j, R_j, and C_j denote the output time, release time, and cycle time of job j, respectively. Since the release time of a job is fixed, shorter cycle time means closer output time, which means faster delivery to the customer. This feature is especially important when market demand fluctuates.

The cycle time of a job can be decomposed as

$$C_j = P_j + W_j \tag{1.2}$$

where P_j and W_j denote the cycle time and total processing time of job j, respectively. The total processing time is fixed. Therefore, unnecessary waiting needs to be avoided to reduce cycle time. That is also the goal of lean production [7]. When a factory is crowded, that is, when the work-in-progress (WIP) level is high, jobs tend to spend long time waiting, resulting in long cycle times.

Jobs of the same product type may have unequal cycle times, since the production conditions when they were manufactured differed. For the same reason, jobs of the same product type in different factories usually have unequal cycle times.

A manufacturing order is usually split into multiple jobs, which may not be released at the same time. The release time of a manufacturing order $\left(R_{M_t}\right)$ is the release time of the first job for the manufacturing order:

$$R_{M_t} = \min_{j \in M_t} R_j \tag{1.3}$$

where M_t indicates manufacturing order t. The cycle times of these jobs are usually unequal. Among them, the latest completed job determines the cycle time of the manufacturing order:

$$C_{M_t} = \max_{j \in M_t} C_j \tag{1.4}$$

The completion time (i.e., makespan) of the manufacturing order is

$$O_{M_t} = \max_{j \in M_t} O_j = \max_{j \in M_t}\left(R_j + C_j\right) \tag{1.5}$$

For a factory, the cycle times of all product types represent its flexibility, controllability, and responsiveness and are therefore crucial for its competitiveness and sustainability [8]. Take a make-to-stock (MTS) factory, for example, where the factory continuously makes products without receiving orders and then tries to sell them. If products have long cycle times, the factory may miss price peaks and can only sell products when market price is lower, thus reducing the profitability and competitiveness of the factory. Conversely, factories with high flexibility and controllability will speed up production (i.e., reduce cycle times) to meet price peaks. A flexible make-to-order (MTO) factory can produce different orders of the same product with different cycle times based on customer requirements and the prices they are willing to pay.

A similar concept is the **step cycle time** of a job (SC_{jk}), which measures the time from when job j is queued for a particular machine until the job completes operation k on the machine (see Fig. 1.1); $k = 1 \sim K$. Furthermore, the **remaining cycle time** (RC_j) is the time required until all operations of a job are completed. Therefore,

$$RC_j = C_j - t \tag{1.6}$$

where t is the current time. Obviously,

$$C_j = \sum_{k=1}^{K} SC_{jk} \tag{1.7}$$

and

$$C_j = \max RC_j \tag{1.8}$$

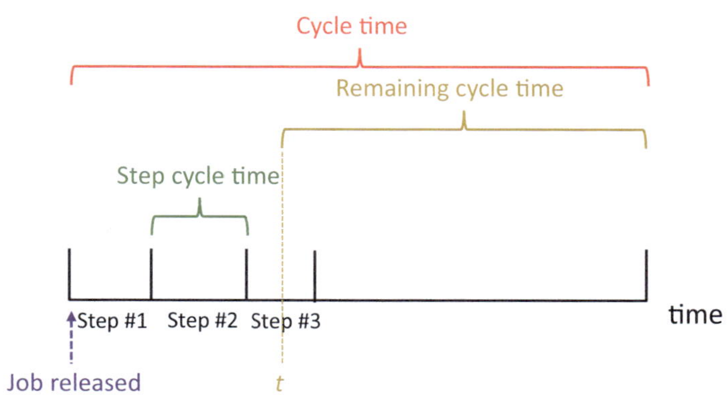

Fig. 1.1 Cycle time, remaining cycle time, and step cycle time

1.2 Cycle Time Management

Cycle time management involves activities such as cycle time forecasting, cycle time modeling and analysis, management support, cycle time control, cycle time reduction, and cycle time learning [9], as illustrated in Fig. 1.2.

The cycle time management process is shown in Fig. 1.3, showing that it is not necessary to perform all six cycle time management activities. In addition, management support can be performed after cycle time forecasting. In addition, cycle time learning can be performed even without management support. Cycle time management has been widely implemented in many manufacturing and service systems, e.g., the semiconductor manufacturing industry. Samsung Electronics Co., Ltd. (SEC), one of the world's largest foundries, implemented a 5-year project called short cycle time and low inventory manufacturing (SLIM) to improve cycle time management [10]. The focus was on how to accelerate the progress of certain wafer lots, that is, shorten their cycle times, to achieve the monthly output target. Another case came from a wafer fabrication plant (fab) that was ramping up its capacity. In this case, short and stable cycle times were a prerequisite for timely equipment qualification and fast yield learning [11], which was especially important for the ramp-up wafer fab. The environment in the wafer fab was characterized by high variability and uncertainty, requiring an approach to properly manage the conflict between short cycle times and rapid throughput growth.

In theory, cycle time management seems more challenging for service systems because long cycle times are often caused by unnecessary waiting, which is mainly caused by inconsistent operations performed or assisted by humans. In addition, the cycle time for providing a service sometimes depends on customer requirements. For example, Chen [12] established a fuzzy just-in-time (JIT) ubiquitous service network, which was an innovative application of mobile commerce, ubiquitous computing, and ambient intelligence. The target user of this system was a traveler who decided the best service location along the planned route. The system then arranged the service when the user needed it so that when the user arrived at the service location, the required service was just ready, thereby providing a better service to the user. In this case, the expected cycle time was equal to the user's travel time. If the user was far away, the service system could slow down the preparation of the required service. Conversely, if the user was approaching, it could speed up the preparation. However, due to the inaccurate user location data, how to determine the JIT service location

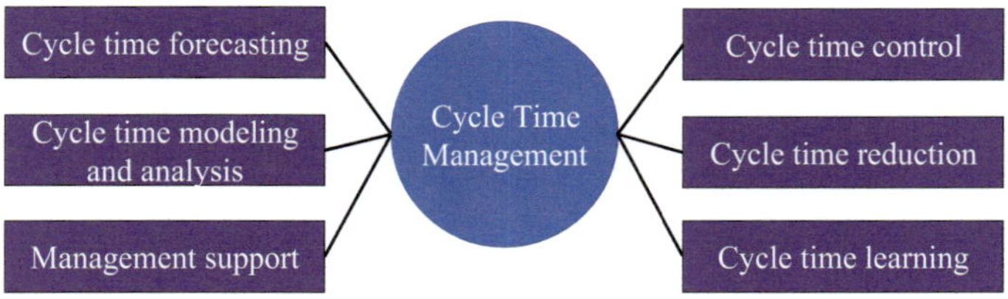

Fig. 1.2 Activities of cycle time management

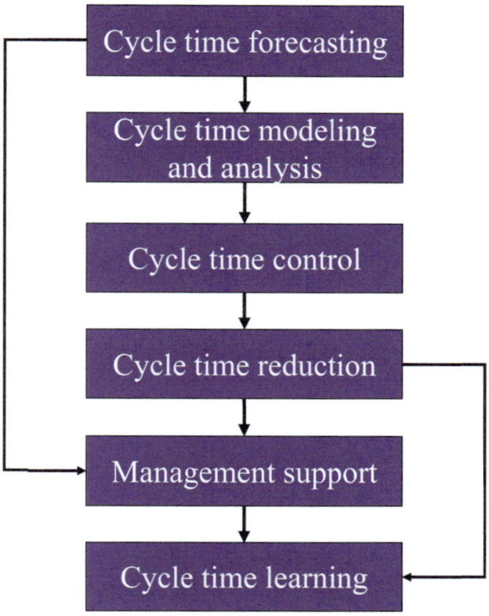

Fig. 1.3 Cycle time management process

and path in the ubiquitous service network was crucial but challenging. To solve this problem, Chen [12] formulated a fuzzy mixed integer-nonlinear programming (FMINLP) problem and proposed a fuzzy version of Dijkstra's algorithm to solve the FMINLP problem.

Therefore, for service systems, cycle time control, that is, controlling the actual cycle time to be exactly equal to the target cycle time, is crucial, while in manufacturing systems, cycle time reduction is more valued by management [13, 14].

1.2.1 Cycle Time Prediction

Cycle time prediction for each job is a critical task to a factory. The results of cycle time prediction also support many management activities, such as internal due date assignment, output projection, ordering decision support, order progress tracking, job scheduling, guiding subsequent operations, and increasing product diversification.

Backus et al. [15] compared three statistical methods for cycle time prediction: clustering (i.e., the average cycle time in each cluster is used as the representative value), K-nearest neighbors (K-NN), and classification and regression trees (CART). They also discussed the advantages and disadvantages of each method. In their experiments, the joint application of clustering and CART achieved the highest prediction accuracy.

Chen [16] proposed a fuzzy neural multi-bucket approach to estimate the cycle times of lots in a wafer fabrication plant (wafer fab) with a dynamic product mix. In this method, the multi-bucket technique took into account the future production plan of the wafer fab. Subsequently, the fuzzy c-means (FCM)-fuzzy backpropagation network (FBPN) method was used to predict the cycle time of each lot in the wafer fab. The buckets obtained in the first stage were used as additional inputs to the FBPN. In this way, the fluctuation of the product mix since the lot was released could be considered when estimating the cycle time of the lot. However, the FBPN was not easy to comprehend and/or communicate [17], which needs to be tackled using explainable artificial intelligence (XAI).

1.2.2 Cycle Time Modeling and Analysis

Cycle time modeling and analysis is to model the cycle time of a job, fit the relationship between job attributes and/or production conditions and the cycle time, and analyze (or explain) the fitted relationship if it is difficult to understand or communicate.

Production simulation is the traditional way to fit the cycle time distribution. Sivakumar and Chong [18] built a discrete event simulation model for a semiconductor backend manufacturing system to fit the cycle time distribution. In addition, they performed production simulation to evaluate the impact of controllable input variables on throughput, cycle time, and cycle time spread. The controllable input variables included lot release policies, lot dispatching rules, the elimination of selected processes, material handling time, setup time, and machine up time.

However, production simulation is time-consuming and subject to many assumptions. In addition, a production simulator can also be used to assess the impact of changing production conditions on the average cycle time and cycle time variation. For example, Chen [19] designed a nonlinear fluctuation smoothing rule for mean cycle time (FSMCT) for job scheduling in a wafer fab to minimize the average cycle time, where the values of two parameters needed to be optimized. To this end, the impact of these two parameters on the average cycle time was evaluated using production simulation. Then, the evaluation results were fitted using the response surface method (RSM).

1.2.3 Management Support

As mentioned previously, cycle time prediction results support many management activities.

According to Tan et al. [20], cycle time prediction results provide an important basis for scheduling control, material procurement, and due date assignment. Many scheduling methods consider cycle time forecasts when prioritizing jobs, such as the fluctuation smoothing rule for cycle time variation (FSVCT) and FSMCT [19]. In addition, the critical ratio (CR) rule will also be more effective if the remaining cycle time of a job can be predicted and used in the rule instead of the remaining processing time [21].

Asghari et al. [22] proposed a system dynamics approach to identify factors that affect the cycle time of material procurement, which was one of the most critical performance indicators for measuring organizational agility. They first defined and modeled subsystems, including internal subsystems such as production and sales processes and external subsystems such as markets and suppliers, and then analyzed the interactions between these subsystems. Subsequently, they established a causal loop and stock-flow model for the problem.

Internal due date assignment is an important management application of cycle time prediction in customer relationship management (CRM) [23]. Chen et al. [23] proposed a hybrid neural network and selective allowance approach for internal due date assignment in a wafer fab. They first constructed a look-ahead self-organization map (SOM)-fuzzy backpropagation network to predict the cycle time of a job. After adding the release time to the predicted cycle time, the completion time of the job was estimated. Subsequently, a selective allowance policy was developed to determine the allowance to be added to the estimated job completion time.

Cycle time management affects supply chain management in the following ways. First, the results of cycle time forecasting provide valuable information for the planning of subsequent processes (including those of downstream factories). Subsequently, if the production processes of jobs go smoothly, that is, their cycle times are properly controlled, then the production plans of subsequent processes and downstream factories are still feasible. Otherwise, they should be notified as soon as possible so that corrective measures can be taken on their production plans at the earliest.

The impact of cycle time management on human resource management is as follows. The step cycle time of an operation is also subject to uncertainty (or variation), and this local uncertainty may or may not affect the control of the cycle time. If it is a problem, workers with a stable work rhythm should be assigned to perform the operation. Otherwise, the selection of workers for this operation is more flexible.

1.2.4 Cycle Time Control

Cycle time control is the process of monitoring the progress of jobs, estimating their remaining cycle times, and formulating and taking actions to speed up or slow down the progress of certain jobs when necessary to meet monthly output targets.

One of the most famous cycle time control methods is the SLIM method proposed by Leachman et al. [10], which speeded up the progress of certain wafer lots to achieve the monthly output target. The cycle times of these wafer lots were thus shortened; however, the cycle times of other wafer lots were inevitably extended. In other words, SLIM is a technique for dynamically changing the priorities of jobs to speed up factory output. Therefore, the SLIM method was a cycle time control method rather than a cycle time reduction method.

Table 1.1 Cycle time control mechanisms for various scheduling objectives

Scheduling objective	Cycle time control mechanism
Mean lateness	Control the progress (cycle times) of jobs which cycle times are relatively short so that they can be completed well in advance of their due dates
Maximum tardiness	Control the progress (cycle times) of the most urgent jobs
Number of tardy jobs	Control the progress (cycle times) of jobs that are more likely to be completed on time
Makespan	Control the progress (cycle time) of the last completed job
Average cycle time	Control the progress (cycle times) of jobs with longer cycle times

Job release and scheduling are perhaps the most common cycle time control mechanisms. There are many job scheduling methods that aim to minimize the makespan (i.e., the last completion time or the cycle time plus release time of the last job), average cycle time, cycle time variation, or their variants, such as shortest processing time (SPT), weighted shortest processing time (WSPT), FSMCT, and FSVCT.

Another equally common cycle time control technique is takt time, which is a lean manufacturing technique that aims to set goals for adjusting the step cycle time of each operation to meet customers' demand rate. JIT [12] is another cycle time control technique in lean manufacturing, where orders are delivered just before their due dates, which is equivalent to controlling the cycle time to be equal to the interval from now to the due date when jobs for the order are released as soon as possible. In contrast, scheduling rules that aim to minimize mean lateness, maximum tardiness, or the number of tardy jobs do not necessarily control the cycle times of all jobs, although they also try to reduce the gap between the completion time and the due date [19]. For example, to minimize the mean lateness, the cycle times of some jobs will be relatively short so that they can be completed well in advance of their due dates. To minimize the maximum tardiness, only the cycle times of the most urgent jobs need to be controlled. To minimize the number of tardy jobs, only the cycle times of some jobs need to be controlled so that they can be completed on time, while the cycle times of other jobs are not controlled and may fall far behind their due dates. All of these rely on different cycle time control mechanisms (see Table 1.1).

1.2.5 Cycle Time Reduction

Cycle time reduction is obviously the most concerned activity in cycle time management. Taifa et al. [24] reviewed cycle time reduction techniques and their effects. The results showed that the most popular cycle time reduction techniques include statistical methods (such as Taguchi method, analysis of variance (ANOVA), and control charts), project management tools (such as critical path method (CPM) and project evaluation and review technique (PERT)), and lean management techniques (such as 5S, value stream mapping (VSM), and line balancing).

Reducing cycle time plays an important role in improving the competitiveness and sustainability of semiconductor manufacturers. However, in the past, cycle time reduction was usually unplanned due to the lack of a systematic and quantitative procedure. To address this problem, Chen [8] established a systematic procedure to plan actions to reduce cycle times in order to improve the competitiveness and sustainability of a semiconductor manufacturer. First, some controllable factors that affect the cycle times of wafer lots were identified. Subsequently, the relationship between the controllable factors and cycle times was fitted using a backpropagation network (BPN). Based on this relationship, actions to reduce cycle times were planned. Before implementing the actions, their feasibility and effectiveness were also evaluated.

Kuo et al. [25] constructed an artificial neural network (ANN) to leverage the value of production and tool data to predict the WIP levels of tool sets to reduce cycle times, taking into account tool specificity and waiting time constraints.

1.2.6 Cycle Time Learning

There are various cycle time learning sources (see Fig. 1.4). First, improving production activities through learning can help reduce cycle times. Since most steps in processing a job require human intervention, learning can be used to shorten the processing times or step cycle times of these steps, thereby shortening the cycle time of the job. In addition, learning across factories or products can help improve the effectiveness of cycle time prediction. If the same product type is produced at two different factories simultaneously, the cycle time at either factory can be used to estimate the cycle time at the other factory. Furthermore, if the manufacturing processes of two product types are similar, the cycle time of one product type can be used as a reference for estimating the cycle time of the other product type.

Job sequencing and scheduling can also be improved by learning from past successes or failures [26], which is another source of cycle time learning. It is sometimes difficult for experienced schedulers to clearly express how they sequence or schedule jobs to shorten cycle times. To address this problem, Gombolay et al. [26] used a variety of machine learning (ML) techniques, including decision tree, K-NN classification, logistic regression (logit), support vector machine (SVM) based on the radial basis function (RBF) kernel, and ANN to learn the scheduling mechanisms adopted by human experts.

Deep neural networks (DNNs) have been used to accurately predict the cycle times of jobs in manufacturing systems. However, the prediction mechanism of a DNN is complex and difficult to communicate. This limits its acceptability (or practicality) in real-world applications. Wang et al. [27] constructed a CART to explain the prediction mechanism of a DNN for job cycle time prediction. The outcome of each branch in the CART model was also replaced by a fuzzy linear regression (FLR) equation

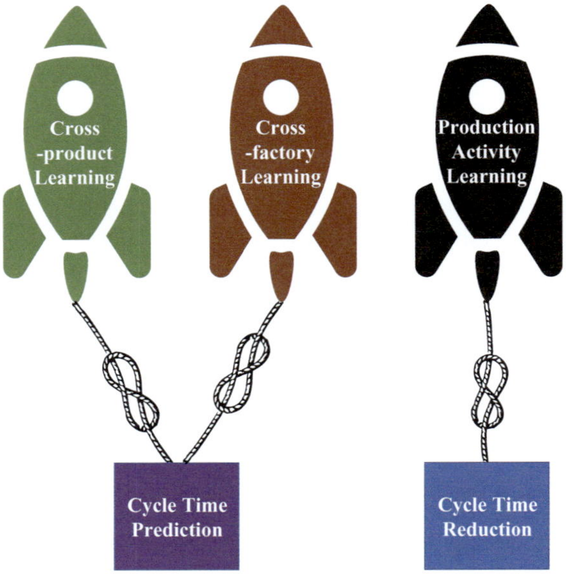

Fig. 1.4 Various cycle time learning sources

for estimating the cycle time range to compensate for its insufficient precision. The results contained some fuzzy decision rules that could be easily learned by human experts, such as "If the lot size is moderate, and the factory utilization is high, and the queue length before the bottleneck is long, then the cycle time is moderate."

Although cycle time management may not be mentioned explicitly in factories, related activities are widely and continuously carried out, especially cycle time forecasting and cycle time reduction, and the importance of these activities has been fully recognized. For example, aiming to improve/shorten cycle time. Kalyar et al. [28] used a questionnaire to evaluate the importance of cultural competitiveness, market turbulence, and knowledge development on cycle time management performance. The survey results showed that cultural competitiveness and market turbulence had a positive impact on cycle time management, but the correlation between knowledge development and cycle time management performance was not significant.

Figure 1.5 shows statistics on recent references to cycle time management activities. The most studied cycle time management activities include cycle time reduction, cycle time modeling and analysis, and management support. The percentages of these three cycle time management activities add up to about 80%.

1.3 Difficulties in Cycle Time Management Manufacturing

Existing cycle time management methods face many challenges. First, in large-scale or complex manufacturing systems, the cycle times of jobs are usually long and highly uncertain, which makes it very difficult to predict the cycle times using

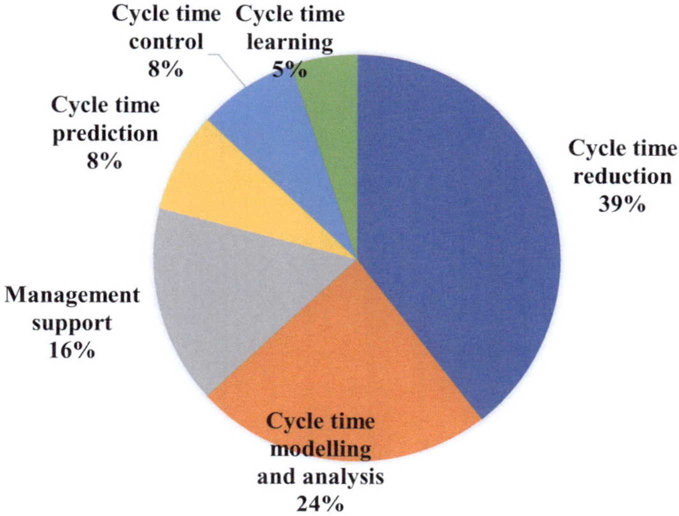

Fig. 1.5 Statistics on recent references to cycle time management activities (*Data source* Google Scholar)

existing methods. In addition, many factors may affect the cycle time of a job, forming a big data problem that exceeds the analytical capabilities of traditional statistical methods.

In addition, while most manufacturing systems seek to reduce cycle time, other management goals may be equally important, such as maximizing throughput or meeting due dates. Therefore, orders are released as soon as they are received, although this practice may increase the WIP level and lengthens cycle time. Other scheduling performances also need to be optimized in addition to average cycle time or cycle time variation.

In practice, a cycle time reduction goal is often set such that cycle time decreases linearly over time. However, this treatment may not be feasible, since the expected cycle time would then drop to zero or even a negative value. In fact, cycle time, like other performance indicators, will improve following a learning process, and the asymptotic value is the total processing time. The cycle time reduction goal should be based on the corresponding cycle time learning model.

Deep learning (DL) methods have been extensively applied to enhance the accuracy of predicting the cycle times of jobs in manufacturing systems. However, the prediction mechanisms of such DL methods are usually complex and difficult to communicate. This limits their acceptability, trust, and practicality in real-world applications. XAI is expected to address this issue.

Furthermore, many improvement activities are closely related to cycle time management. For example, in lean manufacturing, avoiding unnecessary waiting directly reduces the average cycle time. Moreover, mapping the step cycle time of each step to the takt time helps minimize cycle time fluctuations. In job scheduling,

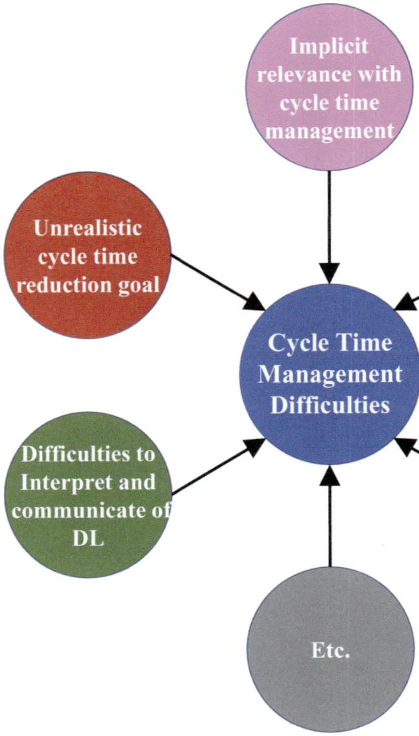

Fig. 1.6 Difficulties in performing cycle time management activities

a common scheduling goal is to minimize the makespan (i.e., the longest completion time), which can be achieved by reducing the cycle time of each job or at least the last completed job. In these efforts, cycle time may not be explicitly mentioned, but its relevance is clearly high. Therefore, researchers on cycle time management should consult a wide range of references to gain useful insights.

Figure 1.6 summarizes the difficulties in performing cycle time management activities.

1.4 Advanced Information Technologies with Applications in Cycle Time Management

Recently, some advanced information technologies have emerged and can be used to overcome the difficulties and further enhance the effectiveness of cycle time management (see Fig. 1.7), such as Industry 4.0, big data analysis, edge computing, cloud and ubiquitous computing, and XAI:

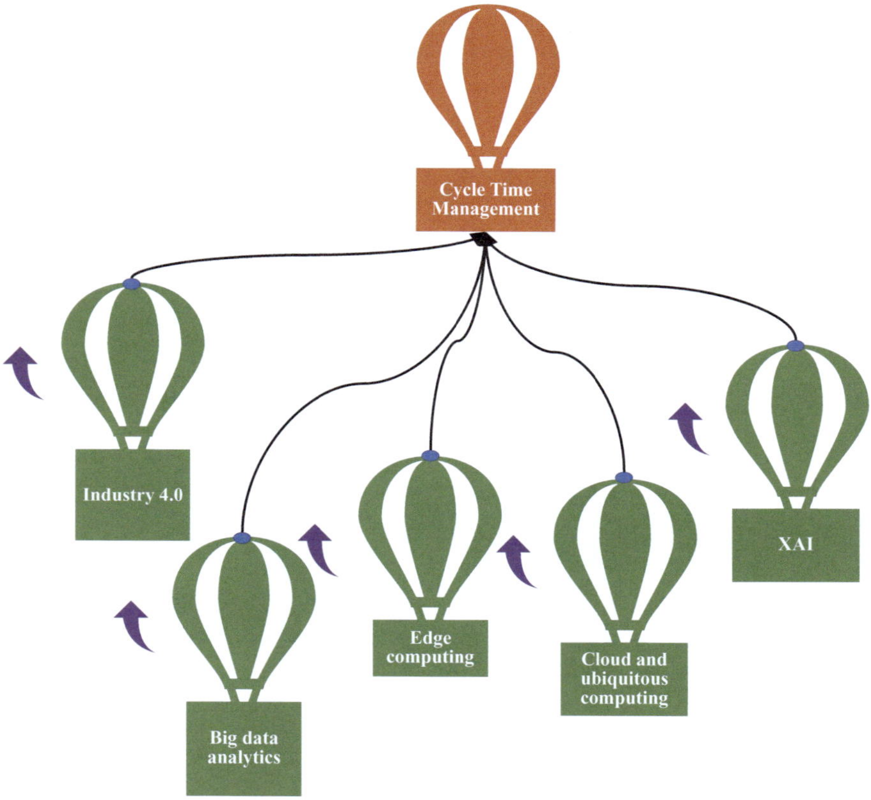

Fig. 1.7 Advanced information technologies and cycle time management

- **Industry 4.0**: Industry 4.0 was introduced by the German government to build smart factories, which is viewed as the ultimate realization of smart manufacturing [29, 30].
- **Big data analytics**: Big data analytics is to examine large amounts of data to uncover hidden patterns, correlations, and other insights [31, 32].
- **Edge computing**: Edge computing is different from traditional cloud computing. It is a new computing paradigm that performs computing at the edge of the network. The core idea of edge computing is to bring computing closer to the source of data [33].
- **Ubiquitous computing**: Ubiquitous manufacturing is an application of ubiquitous computing in the manufacturing sector to enable convenient, on-demand network access to a shared pool of configurable manufacturing resources, including computer software and hardware, equipment, and manufacturing capability [34].
- **XAI**: XAI is to generate more explainable ML models, while maintaining a high level of learning performance (prediction or classification accuracy) [35, 36].

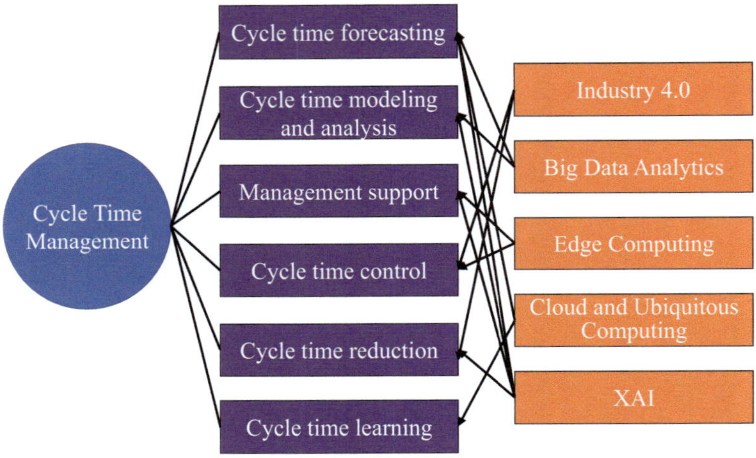

Fig. 1.8 Applications of advanced information technologies for each cycle time management activity

Industry 4.0 technologies such as the industrial Internet of things (IIoT) have been applied to achieve more precise cycle time control of assembly lines. Cyber-physical systems (CPS) or digital twins are also constructed to assist with cycle time modeling and analysis. Big data technologies have been applied to process cycle time-related data for feature selection and dimensionality reduction, so that DL can be used to achieve accurate cycle time predictions, based on which edge computing technologies can be used to assist managerial activities such as internal due date assignment. XAI techniques can then be employed to enhance the explainability and trust of these advanced information technology applications (see Fig. 1.8).

Ruppert and Abonyi [37] developed an IIoT-based solution to ensure the real-time connectivity between products and assembly lines. They also proposed a dynamic cycle time setting algorithm that took into account the different complexities of products and real-time production conditions provided by smart wireless sensors and indoor positioning systems (IPSs).

Chen and Lin [38] designed a cloud and edge computing scheme for internal due date assignment, in which a sales representative visited a customer and negotiated with him/her the due date of a possible order. In the past, such tasks were accomplished by querying a DL system on the backend or cloud server, which was affected by difficulties such as cloud computing software licensing, network speed, and bandwidth. To overcome these difficulties, in the cloud and edge computing scheme, a possible order was broken down into lots that are released based on daily quotas. A DNN was then constructed on the cloud server to estimate the output time of each lot, thereby obtaining the completion time of the possible order for internal due date assignment. A random forest (RF) was also built to approximate the trained DNN for internal due date assignment using edge computing.

Chen and Wang [39] proposed a hybrid big data analytics and Industry 4.0 (BD-I4) approach to enhancing the effectiveness of cycle time range estimation for jobs in a factory. In the BD-I4 approach, each expert constructed a fuzzy DNN to estimate the cycle time range of a job, an application of big data analytics and deep learning. Subsequently, the fuzzy weighted intersection (FWI) operator was applied to aggregate the cycle times estimated by all experts such that their unequal authority levels could be considered, an application of Industry 4.0. After applying the BD-I4 approach to a real case, the estimation precision was improved by up to 72%, suggesting that instead of relying on a single expert, collaboration among multiple experts might be more effective and efficient.

Kocsi et al. [40] developed an advanced and effective real-time production scheduling decision support system for high-mix low-volume (HMLV) manufacturing systems using Industry 4.0 technologies. They also applied robotic process automation and several advanced scheduling techniques controlled by the analytic hierarchy process (AHP). The real-time production scheduling decision support system aimed to minimize the makespan. In addition, they also adopted a combination of multi-agent optimization and genetic algorithm (GA) to mitigate the potential risks in the production process.

To predict the cycle times of wafer lots, Wang and Zhang [5] established a feature selection process based on conditional mutual information to select key features through dimensionality reduction without any prior knowledge. In addition, to further improve the efficiency of processing massive data, a concurrent prediction mode was adopted to predict the cycle times of wafer lots in parallel.

The traditional method for fitting cycle time distribution is simulation, but this method is time-consuming and requires a lot of assumptions. Tan et al. [20] proposed a big data analysis method to achieve the same goal. They built a density peak-based radial basis function network for cycle time prediction based on the Apache Hadoop software library, which could be distributed across computer clusters to process large cycle time datasets.

In real-world manufacturing, the frequent occurrence of abnormal disturbances can lead to the deviations from the scheduling results, affecting the accuracy and reliability of the scheduling plan. To address this problem, Wang et al. [41] proposed a real-time digital-twin flexible workshop scheduling method based on edge computing to reduce the impact of abnormal disturbances on the deviations from the scheduling results. The method optimized three scheduling objectives: makespan, workload, and total energy consumption.

DL methods have been widely used to improve the effectiveness of cycle time prediction. However, the prediction mechanisms are often difficult to understand and communicate, which hinders their acceptability and practicality. To address this issue, XAI techniques have been adopted to approximate the prediction process and/or results using simpler decision rules. However, these decision rules may not be consistent with domain knowledge, which limits their practicality. To address this issue, Chen and Wang [17] proposed an improved random forest-based incremental interpretation (RFII) method, which used an RF to fit the input–output relationship of the DL application through decision rules. Then, the domain knowledge of experts was extracted and the decision rules were checked against such knowledge. Decision rules with poor consistency with domain knowledge were excluded. Subsequently, the remaining decision rules were aggregated in a way that was convenient for users to interpret.

A special case of cycle time management is cycle time management in a supply chain. Since the manufacture of a product or the provision of a service requires the participation of many participants in the supply chain, the total cycle time needs to be compressed, namely, the cycle time of each participant needs to be shortened, and the information sharing between supply chain participants also needs to be efficient. Che and Chiang [42] aimed to address supplier selection, supply chain planning, and production and distribution planning for a supply chain simultaneously. For supplier selection, they applied AHP with rough sets theory to establish a supplier purchasing value rating system. Subsequently, for supply chain planning, they established a cycle time prediction procedure to effectively predict the total cycle time through a collaborative supply chain. Assembly-line balancing techniques were then employed to smooth the production and distribution in the supply chain.

1.5 Organization of This Book

For factories, the cycle times of all products reflect their agility, controllability, and responsiveness, and is therefore crucial for their competitiveness [43, 44] and sustainability [45]. Cycle time management can improve the competitiveness and sustainability of factories through cycle time prediction, cycle time modeling and analysis, management support, cycle time control, cycle time reduction, and cycle time learning. Among them, cycle time reduction is regarded as a vital and urgent task for all types of factories. In recent years, some advanced information technologies, such as Industry 4.0 [46], big data analytics, edge computing, cloud computing and ubiquitous computing, and XAI [47, 48], have emerged to further enhance the effectiveness of cycle time management. This book aims to provide technical details about cycle time management and explore the applications of advanced information technologies in this field. These details should be of great interest to researchers, practicing managers, and engineers in the fields of lean manufacturing, industrial and mechanical engineering, operations management, production control, competitiveness management, and information engineering and management.

In specific, the outline of the present book is structured as follows.

This chapter first defines the basic concepts, such as cycle time, step cycle time, and remaining cycle time, and discusses the relationship between these times. Subsequently, cycle time management is defined and its six activities are introduced: cycle time prediction, cycle time modeling and analysis, management support, cycle time control, cycle time reduction, and cycle time learning. Relevant cases or references are reviewed to illustrate the importance and implementation of each activity. From the review results, the most studied cycle time management activities include cycle time reduction, cycle time modeling and analysis, and management support. In addition, difficulties faced by cycle time management are also summarized. Recently, some advanced information technologies have emerged that can be used to overcome the difficulties of cycle time management and improve its effectiveness: Industry 4.0, big data analytics, edge computing, cloud computing and ubiquitous computing, XAI, etc. How these advanced information technologies can help improve the effectiveness of cycle time management are discussed. References supporting these potential applications are also reviewed.

Chapter 2—Cycle Time Prediction—first highlights the importance of cycle time prediction by mentioning the management activities supported by cycle time prediction results. Existing cycle time prediction methods are then divided into six categories: statistical methods, production simulation, machine learning or deep learning methods, case-based reasoning, fuzzy modeling methods, and hybrid methods. For each type of methods, the mathematical background behind it is first explained, and then numerical examples and program codes are provided. In recent years, some advanced information technologies have emerged that can be used to enhance the performance of cycle time prediction: Industry 4.0, big data analysis, edge computing, cloud computing and ubiquitous computing, XAI, etc. How these advanced information technologies cope with the challenges faced by traditional cycle time prediction methods, and some applications of these advanced information technologies in cycle time prediction are introduced. Numerical examples and program codes are given as well.

Chapter 3—Cycle Time Modeling and Analysis—starts by defining the concept of cycle time modeling and analysis. Various cycle time modeling and analysis techniques and tools are then introduced. First, value stream mapping is applied to identify the operations of a job with the longest waiting times and should be improved first. Subsequently, after training the cycle time prediction model using machine learning or deep learning methods, a causal cycle time relationship analysis is performed to evaluate the impact of each job attribute on the cycle time forecast of a job, so as to find out the most important job attributes for the job. To this end, Shapely analysis and related tools are introduced with numerical examples and program codes. Subsequently, to explain and communicate the cycle time prediction mechanism and results, several XAI techniques and tools can be applied. First, for cycle time prediction problems involving big data, job classification is usually performed. Therefore, before predicting the cycle times, jobs need to be classified. To this end, traditional and XAI tools for explaining the job classification process and results are reviewed. Seven requirements that need to be met for excellent explanations are also listed. Subsequently, existing and XAI techniques and tools are applied to explain the cycle time prediction process and results, such as random forest-based incremental explanation. In addition, a systematic procedure is also established to confirm whether a trained cycle time prediction model conforms to domain knowledge, and on this basis, the improved random forest-based incremental explanation technique is proposed.

Chapter 4—Management Support—discusses the results of cycle time prediction, analysis, and modeling which can support various management activities, such as internal due date assignment, output projection, job sequencing and scheduling, competitiveness enhancement, customer relationship management, human resource management, and supply chain management. This chapter describes how to apply the results of cycle time prediction, analysis, and modeling to these management activities. Two management activities are especially focused on: internal due date assignment and output projection. Internal due date assignment refers to determining the date on which the manufacturing system can complete and deliver a potential order. The internal due date of a potential order can be determined by adding a certain allowance to the estimated completion time of the potential order. Advanced information technologies, such as cloud and edge computing, can also be applied to internal due date assignment. Therefore, a hybrid cloud and edge computing approach to enhance the effectiveness of internal due date assignment in advanced manufacturing systems is reviewed. In addition, imprecise output projection forecasts can lead to delivery delays, loss of potential orders, or unnecessary inventory accumulation, which is a critical challenge for manufacturing systems to overcome. Therefore, various machine learning or deep learning methods have been applied to output projection. Advanced information technologies, such as Industry 4.0 and explainable artificial intelligence, can also be applied to enhance the effectiveness and interpretability of output projection.

Chapter 5—Cycle Time Reduction—discusses that, for many manufacturing systems, cycle time reduction is obviously the key to improving competitiveness. Therefore, cycle time reduction is regarded as the most critical and urgent task in production control. This chapter first summarizes the benefits of cycle time reduction on manufacturing systems from existing references and practices, such as reducing work-in-process levels, improving productivity, reducing unit costs, responding to market demand changes, and improving yield. This chapter also points out that cycle time reduction does not necessarily mean speeding up the execution of all operations. In addition, the cycle time reduction target of a product can be decomposed into the targets of its individual components based on the results of the bill of materials or material requirement planning. Subsequently, some methods for reducing cycle time are introduced, including more efficient job scheduling, yield improvement, controlling production conditions, 3D printing, leveraging cloud manufacturing capabilities, pull manufacturing, value stream mapping, and lean six sigma. Obviously, cycle time can be reduced in all aspects of system planning, control, and management. Finally, this chapter also provides numerical examples to illustrate the application of these cycle time reduction methods and discusses their potential effects.

References

1. T. Chen, C.-W. Lin, Y.-C. Lin, A fuzzy collaborative forecasting approach based on XAI applications for cycle time range estimation. Appl. Soft Comput. **151**, 111122 (2024)
2. L. Frady, Flow time versus cycle time: what's the difference? https://www.isixsigma.com/methodology/flow-time-vs-cycle-time-whats-the-difference/
3. R.M. Ali, A.M. Deif, Dynamic lean assessment for takt time implementation. Procedia CIRP **17**, 577–581 (2014)

4. T.C.T. Chen, Y.C. Wang, AI applications to pull production, JIT, and production leveling, in *Artificial Intelligence and Lean Manufacturing* (2022), pp. 55–74
5. J. Wang, J. Zhang, Big data analytics for forecasting cycle time in semiconductor wafer fabrication system. Int. J. Prod. Res. **54**(23), 7231–7244 (2016)
6. AeroCorner.com, The 9-day process building a Boeing 737 MAX (2024). https://aerocorner.com/blog/how-airplanes-are-made/#the-9-day-process-building-a-boeing-737-max
7. T.C.T. Chen, Y.C. Wang, Basics in lean management, in *Artificial Intelligence and Lean Manufacturing* (2022), pp. 1–12
8. T. Chen, A systematic cycle time reduction procedure for enhancing the competitiveness and sustainability of a semiconductor manufacturer. Sustainability **5**(11), 4637–4652 (2013)
9. SixSigma.us, Understanding cycle time: definition, calculation, and importance (2024). https://www.6sigma.us/manufacturing/cycle-time/
10. R.C. Leachman, J. Kang, V. Lin, SLIM: short cycle time and low inventory in manufacturing at Samsung electronics. Interfaces **32**(1), 61–77 (2002)
11. M. Haller, A. Peikert, J. Thoma, Cycle time management during production ramp-up. Robot. Comput.-Integr. Manuf. **19**(1–2), 183–188 (2003)
12. T. Chen, Creating a just-in-time location-aware service using fuzzy logic. Appl. Spat. Anal. Policy **26**(9), 287–307 (2016)
13. M.M. Amini, D. Retzlaff-Roberts, C.C. Bienstock, Designing a reverse logistics operation for short cycle time repair services. Int. J. Prod. Econ. **96**(3), 367–380 (2005)
14. T. Chen, A fuzzy integer-nonlinear programming approach for creating a flexible just-in-time location-aware service in a mobile environment. Appl. Soft Comput. **38**, 805–816 (2016)
15. P. Backus, M. Janakiram, S. Mowzoon, C. Runger, A. Bhargava, Factory cycle-time prediction with a data-mining approach. IEEE Trans. Semicond. Manuf. **19**(2), 252–258 (2006)
16. T. Chen, A fuzzy-neural and multiple-bucket approach for estimating lot cycle time in a wafer fab with dynamic product mix. Comput. Ind. Eng. **55**(2), 423–438 (2008)
17. T. Chen, Y.C. Wang, A modified random forest incremental interpretation method for explaining artificial and deep neural networks in cycle time prediction. Decis. Analytics J. **7**, 100226 (2023)
18. A.I. Sivakumar, C.S. Chong, A simulation based analysis of cycle time distribution, and throughput in semiconductor backend manufacturing. Comput. Ind. **45**(1), 59–78 (2001)
19. T. Chen, An effective dispatching rule for bi-objective job scheduling in a wafer fabrication factory—considering the average cycle time and the maximum lateness. Int. J. Adv. Manuf. Technol. **67**(5–8), 1281–1295 (2013)
20. X. Tan, L. Xing, Z. Cai, G. Wang, Analysis of production cycle-time distribution with a big-data approach. J. Intell. Manuf. **31**, 1889–1897 (2020)
21. T.C.T. Chen, XAI applications in job sequencing and scheduling, in *Explainable and Customizable Job Sequencing and Scheduling: Advancing Production Control and Management with XAI* (2025), pp. 37–71
22. M. Asghari, F. Rostambeygi, S.H. Hosseini, M. Motamedi, An analysis of influential factors on procurement cycle time using a system dynamics approach. J. Syst. Thinking Pract. **1**(3), 72–91 (2022)
23. T. Chen, A. Jeang, Y.C. Wang, A hybrid neural network and selective allowance approach for internal due date assignment in a wafer fabrication plant. Int. J. Adv. Manuf. Technol. **36**, 570–581 (2008)
24. I. Taifa, T. Vhora, Cycle time reduction for productivity improvement in the manufacturing industry. J. Ind. Eng. Manage. Stud. **6**(2), 147–164 (2019)
25. C.J. Kuo, C.F. Chien, J.D. Chen, Manufacturing intelligence to exploit the value of production and tool data to reduce cycle time. IEEE Trans. Autom. Sci. Eng. **8**(1), 103–111 (2010)
26. M. Gombolay, R. Jensen, J. Stigile, S.H. Son, J. Shah, Apprenticeship scheduling: learning to schedule from human experts, in *International Joint Conferences on Artificial Intelligence* (2016), pp. 1–9
27. Y.C. Wang, T. Chen, M.C. Chiu, An explainable deep-learning approach for job cycle time prediction. Decis. Analytics J. **6**, 100153 (2023)

28. M.N. Kalyar, H.M. Sabir, I. Shafique, Improving cycle time performance: the role of market turbulence and cultural competitiveness. Prod. Manuf. Res. **1**(1), 3–18 (2013)
29. C. Bai, P. Dallasega, G. Orzes, J. Sarkis, Industry 4.0 technologies assessment: a sustainability perspective. Int. J. Prod. Econ. **229**, 107776 (2020)
30. T.C.T. Chen, Industry 4.0 for semiconductor manufacturing, in *Production Planning and Control in Semiconductor Manufacturing: Big Data Analytics and Industry 4.0 Applications* (2022), pp. 21–40
31. SAS Institute, Big data analytics: what it is and why it matters (2021). https://www.sas.com/en_us/insights/analytics/big-data-analytics.html
32. T.C.T. Chen, Big data analytics for semiconductor manufacturing, in *Production Planning and Control in Semiconductor Manufacturing: Big Data Analytics and Industry 4.0 Applications* (2022), pp. 1–19
33. K. Cao, Y. Liu, G. Meng, Q. Sun, An overview on edge computing research. IEEE Access **8**, 85714–85728 (2020)
34. T. Chen, H.R. Tsai, Ubiquitous manufacturing: current practices, challenges, and opportunities. Robot. Comput.-Integr. Manuf. **45**, 126–132 (2017)
35. D. Gunning, D. Aha, DARPA's explainable artificial intelligence (XAI) program. AI Mag. **40**(2), 44–58 (2019)
36. T.C.T. Chen, Explainable artificial intelligence (XAI), in *Explainable and Customizable Job Sequencing and Scheduling: Advancing Production Control and Management with XAI* (2025), pp. 1–20
37. T. Ruppert, J. Abonyi, Industrial internet of things based cycle time control of assembly lines, in *2018 IEEE International Conference on Future IoT Technologies* (2018), pp. 1–4
38. T.C.T. Chen, C.W. Lin, Feasibility of a clouds and edge computing scheme for internal due date assignment in advanced manufacturing systems. Int. J. Adv. Manuf. Technol. **136**(5), 2893–2908 (2025)
39. T. Chen, Y.C. Wang, Hybrid big data analytics and industry 4.0 approach to projecting cycle time ranges. Int. J. Adv. Manuf. Technol. **120**(1), 279–295 (2022)
40. B. Kocsi, M.M. Matonya, L.P. Pusztai, I. Budai, Real-time decision-support system for high-mix low-volume production scheduling in industry 4.0. Processes **8**(8), 912 (2020)
41. J. Wang, Y. Liu, S. Ren, C. Wang, S. Ma, Edge computing-based real-time scheduling for digital twin flexible job shop with variable time window. Robot. Comput.-Integr. Manuf. **79**, 102435 (2023)
42. Z.H. Che, T.A. Chiang, Designing a collaborative supply-chain plan using the analytic hierarchy process and genetic algorithm with cycle-time estimation. Int. J. Prod. Res. **50**(16), 4426–4443 (2012)
43. T. Chen, Z. Mikoláš, Y.-C. Wang, Competitiveness assessment and enhancement for virtual organisations. Int. J. Technol. Manage. **70**(1), 1–3 (2016)
44. A. Sapkauskiene, S. Leitoniene, The concept of time-based competition in the context of management theory. Eng. Econ. **21**(2), 205–213 (2010)
45. T. Chen, Y.-C. Wang, Evaluating sustainable advantages in productivity with a systematic procedure. Int. J. Adv. Manuf. Technol. **87**, 1435–1442 (2016)
46. T. Chen, Y.-C. Wang, C.-W. Lin, Enhancing the effectiveness of output projection in wafer fabrication using an industry 4.0 and XAI approach. Int. J. Adv. Manuf. Technol. **134**, 113–125 (2024)
47. T. Wagner, C. Herrmann, S. Thiede, Industry 4.0 impacts on lean production systems. Procedia CIRP **63**, 125–131 (2017)
48. T.C.T. Chen, C.W. Lin, A hybrid XAI-FCI approach for job cycle time range estimation: wafer fabrication as an example. Oper. Res. Int. J. **25**(2), 1–29 (2025)

Chapter 2
Cycle Time Prediction

Abstract This chapter first highlights the importance of cycle time prediction by mentioning the management activities supported by cycle time prediction results. Existing cycle time prediction methods are then divided into six categories: statistical methods, production simulation, machine learning or deep learning methods, case-based reasoning, fuzzy modeling methods, and hybrid methods. For each type of methods, the mathematical background behind it is first explained, and then numerical examples and program codes are provided. In recent years, some advanced information technologies have emerged that can be used to enhance the performance of cycle time prediction: Industry 4.0, big data analysis, edge computing, cloud computing and ubiquitous computing, explainable artificial intelligence, etc. How these advanced information technologies cope with the challenges faced by traditional cycle time prediction methods, and some applications of these advanced information technologies in cycle time prediction are introduced. Numerical examples and program codes are given as well.

Keywords Cycle time prediction · Machine learning · Deep learning · Industry 4.0 · Big data analytics

2.1 Cycle Time Prediction: Importance and Methods

Predicting the cycle time for each job in a factory is critical not only to the factory itself, but also to customers, as the results of cycle time prediction support many management activities, including internal due date assignment, output projection, ordering decision support, order progress tracking, job scheduling, guiding subsequent operations, increasing product diversification, etc. [1, 2] (see Fig. 2.1).

Traditional cycle time prediction methods can be divided into six categories: statistical methods, production simulation, machine learning (ML) or deep learning (DL) methods, case-based reasoning (CBR), fuzzy modeling methods, and hybrid methods [1] (see Fig. 2.2). Traditional cycle time prediction methods can benefit from applying emerging IT technologies such as Industry 4.0, big data analytics, edge

T.-C. T. Chen, *Advances in Cycle Time Management*,
https://doi.org/10.1007/978-3-032-06911-5_2

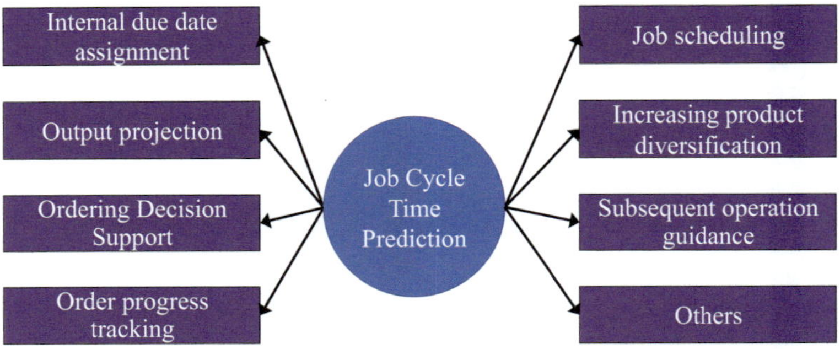

Fig. 2.1 Management activities supported by cycle time prediction

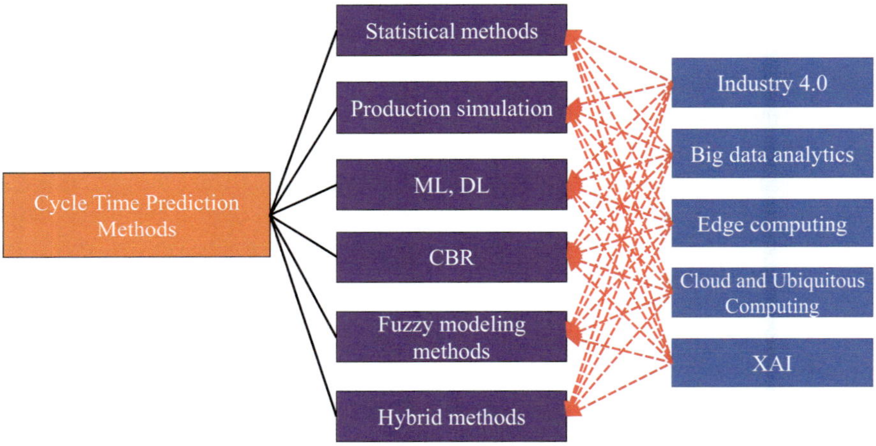

Fig. 2.2 Categories of cycle time prediction methods

computing, cloud and ubiquitous computing, and explainable artificial intelligence (XAI).

2.2 Statistical Methods for Cycle Time Prediction

2.2.1 Historical Statistics, Time Series Analysis

Let the cycle time forecast and actual value of job j be denoted by $\hat{C}_j$ and C_j, respectively. When historical cycle time data has been collected over a period of time, the average cycle time for each product type can be calculated. Then, the cycle time of a new job can be estimated by the average cycle time for the product type of the job in a straightforward way:

$$\hat{C}_j = \frac{\sum_{k=1}^{n} C_k}{n}; j > n \tag{2.1}$$

If there are several product types in the factory, the **cycle time multiplier** (CTM) can be derived as follows:

$$\text{CTM} = \sum_{k=1}^{n} \frac{C_k}{TP_k} \tag{2.2}$$

where TP_k is the total processing time of job k. Then, the cycle time of job j can be predicted as follows:

$$\hat{C}_j = \text{CTM} \cdot TP_j; j > n \tag{2.3}$$

The cycle time forecasting performance (accuracy) can be measured in terms of mean absolute error (MAE), mean arctangent absolute percentage error (MAAPE), mean absolute percentage error (MAPE), root mean squared error (RMSE), and the coefficient of determination $\left(R^2\right)$:

$$\text{MAE} = \frac{\sum_{j=1}^{n} \left|C_j - \hat{C}_j\right|}{n} \tag{2.4}$$

$$\text{MAPE} = \frac{1}{n} \sum_{j=1}^{n} \frac{\left|C_j \_ \hat{C}_j\right|}{C_j} \tag{2.5}$$

$$\text{MAAPE} = \frac{1}{n} \sum_{j=1}^{n} \frac{\left|C_j \_ \hat{C}_j\right|}{C_j} tan^{-1} \frac{\left|C_j \_ \hat{C}_j\right|}{C_j} \tag{2.6}$$

$$\text{RMSE} = \sqrt{\frac{\sum_{j=1}^{n} (C_j - \hat{C}_j)^2}{n}} \tag{2.7}$$

$$R^2 = 1 - \frac{\sum_{j=1}^{n} (C_j - \hat{C}_j)^2}{\sum_{j=1}^{n} (C_j - \overline{C})^2} \tag{2.8}$$

Among them, MAAPE is to avoid the issue of dividing by zero when the actual value is zero (very small), which is likely to happen if C_j is small or is normalized so that it may be close to 0. Another possible case is when building an artificial neural network (ANN) or deep neural network (DNN) to predict the cycle time of a job, the logistic sigmoid function is a common activation function. The output of this function is between 0 and 1. Therefore, the inputs and output of the ANN are usually normalized values, as are the actual values (i.e., cycle times), which are also subject to the same problem and MAAPE needs to be adopted.

Table 2.1 Cycle time series containing 200 jobs

j	C_j (hrs)
1	677
2	687
3	687
…	
200	1373

If jobs for the same product type are manufactured in the factory continuously, their cycle times can be considered as a time series. In this case, time series forecasting methods, such as moving average, exponential smoothing, and **autoregressive integrated moving average model (ARIMA)** [3], can be applied:

$$\left(1-\sum_{i=1}^{p}\phi_i L^i\right)(1-L)^d C_j=\left(1+\sum_{i=1}^{q}\theta_i L^i\right)\varepsilon_j; j>n \tag{2.9}$$

where p is the order (i.e., the number of time lags) of the autoregressive model; d is the degree of differencing (i.e., the number of times to subtract past values from the data); and q is the order of the moving average model. Briefly,

$$\hat{C}_j=\text{ARIMA}(\{C_k\}; p, d, q); j>n \tag{2.10}$$

Example 2.1 The cycle times of 200 jobs for the same product type have been collected, as presented in Table 2.1. The data of the first 150 jobs are used as the training data, while the remaining data are reserved for testing. ARIMA is applied to predict the cycle times of jobs in test data, for which the MATLAB code is provided in Fig. 2.3. Then, the cycle time forecasting performance is evaluated as

$$\text{MAE}=18.7\ (\text{hrs})$$
$$\text{MAPE}=1.7\%$$
$$\text{RMSE}=20.7\ (\text{hrs})$$
$$R^2=0.98$$

2.2.2 *Multiple Linear Regression, Principal Component Regression*

The cycle time of a job can be predicted based on its L attributes (including job size, total processing time, the production conditions at the release time, etc.) using **multiple linear regression (MLR)** [4] as

```
cd F:\ACTM
example1=readtable("example1.xlsx");
example1=table2array(example1);
train_y=example1(1:150, 2:2);
test_y=example1(151:200, 2:2);

d=0;
data1=train_y;
if adftest(train_y)==0 || kpsstest(train_y)==1
  dtrain_y=diff(train_y);
  data1= dtrain_y;
  d=1;
end
if adftest(dtrain_y)==0 || kpsstest(dtrain_y)==1
  ddtrain_y=diff(dtrain_y);
  data1= ddtrain_y;
  d=2;
end

data1=reshape(data1, length(data1), 1);
p_max=4;
q_max=4;
logL_all=zeros(p_max+1, q_max+1);
p_and_q=zeros(p_max+1, q_max+1);
for p=0:p_max
  for q=0:q_max
    model=arima(p, d, q);
    [fit, ~, logL]=estimate(model, data1);
    logL_all(p+1, q+1)=logL;
    p_and_q(p+1, q+1)=p+q;
  end
end
logL_all=reshape(logL_all, (p_max+1)*(q_max+1), 1);
p_and_q=reshape(p_and_q, (p_max+1)*(q_max+1), 1);
m2=length(data1);
[aic, bic]=aicbic(logL_all, p_and_q+1, m2);
aic0=reshape(aic, (p_max+1), (q_max+1));
bic0=reshape(bic, (p_max+1), (q_max+1));

aic1=min(aic0(:));
index1=(aic1==aic0);
[p_all q_all]=meshgrid(0:p_max, 0:q_max);
p0=p_all(index1);
q0=q_all(index1);

aic2=min(bic0(:));
index1=(aic2==bic0);
[p_all q_all]=meshgrid(0:p_max, 0:q_max);
p1=p_all(index1);
q1=q_all(index1);

if p0^2+q0^2>p1^2+q1^2
  p=p1;
  q=q1;
else
  p=p0;
  q=q0;
end

Mdl=arima(p, d, q);
fit1=estimate(Mdl, data1);
[test_o, var_test_o]=forecast(fit1, 50, 'Y0', train_y);
test_rmse=mean((test_y-test_o).^2)^0.5;
test_mae=mean(abs(test_y-test_o));
test_mape=mean(abs(test_y-test_o)./test_y);
test_R2=1-sum((test_y-test_o).^2)/(std(test_y)^2*(50-1));
```

Fig. 2.3 MATLAB codes for implementing ARIMA

$$\hat{C}_j = \beta_0 + \sum_{l=1}^{L} \beta_l x_{jl}; j > n \tag{2.11}$$

where

$$\boldsymbol{\beta} = (\mathbf{X}'\mathbf{X})^{-1}\mathbf{X}'\mathbf{C} \tag{2.12}$$

in which

$$\boldsymbol{\beta} = \left[\beta_0 \ \beta_1 \ \ldots \ \beta_L \right]^T \tag{2.13}$$

$$\mathbf{X} = \begin{bmatrix} 1 \ x_{11} \ \ldots x_{1L} \\ \ldots\ldots \quad \ldots \\ 1 \ x_{n1} \ \ldots x_{nL} \end{bmatrix} \tag{2.14}$$

$$\mathbf{C} = \left[C_1 \ C_2 \ \ldots \ C_n \right]^T \tag{2.15}$$

Example 2.2 Cycle time-related data have been collected for 120 jobs (see Table 2.2), in which the cycle time of job *j*, C_j, is to be predicted based on its six attributes, x_{j1} ~ x_{j6}, using an MLR equation.

The data of the first 90 jobs are used to fit the MLR as

$$\hat{C}_j = -826 + 9.21x_{j1} + 2.00x_{j2} + 1.83x_{j3} - 1.89x_{j4} + 0.13x_{j5} + 354.23x_{j6}$$

Then, the fitted MLR equation is applied to test data that contain the data of the remaining 30 job. Then, the forecasting performance is evaluated as

$$\text{MAE} = 105 \text{ (hrs)}$$
$$\text{MAPE} = 8.2\%$$
$$\text{RMSE} = 121 \text{ (hrs)}$$
$$R^2 = 0.60$$

Table 2.2 Cycle time-related data of 120 jobs

j	x_{j1}	x_{j2}	x_{j3}	x_{j4}	x_{j5}	x_{j6} (%)	C_j (hrs)
1	24	1223	158	807	99	84.2	953
2	23	1225	164	665	142	94.8	1248
3	25	1232	154	718	373	88.4	1299
…							
120	22	1319	159	777	326	88.8	1285

In Eq. (2.11), when there are many job attributes, i.e., a big data problem, calculating the inverse of $\mathbf{X}'\mathbf{X}$ becomes challenging. To address this issue, big data analytics, such as dimensionality reduction techniques, can be applied. **Principal component analysis (PCA)** is among the most prevalent dimensionality reduction techniques by eliminating the multicollinearity between job attributes [5]:

$$Z_p = \sum_{l=1}^{L} \phi_{lp} x_l; \, p = 1 \sim L \tag{2.16}$$

$$\sum_{l=1}^{L} \phi_{lp}^2 = 1; \, p = 1 \sim L \tag{2.17}$$

where Z_p is the p-th principal component and $[\phi_{lp}]$ is its loading vector with components perpendicular to each other. Then, the cycle time of job j can be predicted based on the principal components instead with a principal component regression (PCR) equation as follows:

$$\hat{C}_j = w_0 + \sum_{p=1}^{P} w_p Z_{jp}; \, j > n \tag{2.18}$$

where P (the number of principal components) is usually much smaller than L (the number of original job attributes).

Example 2.3 PCA is applied to remove the multicollinearity between job attributes in Example 2.2, for which the MATLAB code is provided in Fig. 2.4. The results are summarized in Table 2.3, which contains the values of six principal components. C_j is now predicted based on the new variables (principal components) with the highest explaining abilities, i.e., $Z_{j1} \sim Z_{j3}$, using a PCR equation as

$$\hat{C}_j = 1225 + 0.08Z_{j1} + 2.65Z_{j2} - 0.97Z_{j3}$$

The forecasting accuracy is evaluated as

$$\text{MAE} = 98 \text{ (hrs)}$$

```
train_x = [24 1223 158 807 99 0.842; …; 25 1389 166 769 297 0.882];
test_x = [23 1335 177 650 30 0.784; …; 22 1319 159 777 326 0.888];
[coeff, score, latent, tsquared, explained, mu] = pca(train_x);
new_train_x = (train_x - mu)*coeff(:, :);
new_test_x = (test_x - mu)*coeff(:, :);
```

Fig. 2.4 MATLAB code for implementing PCA

Table 2.3 PCA results

j	Z_{j1}	Z_{j2}	Z_{j3}	Z_{j4}	Z_{j5}	Z_{j6}
1	− 157.87	− 85.29	47.01	− 11.52	1.16	− 0.06
2	− 112.33	− 26.47	− 81.45	− 7.14	− 0.01	0.05
3	117.31	− 35.37	− 22.77	− 15.17	1.98	− 0.02
…						
120	66.21	20.25	64.13	− 10.04	− 1.07	0.00

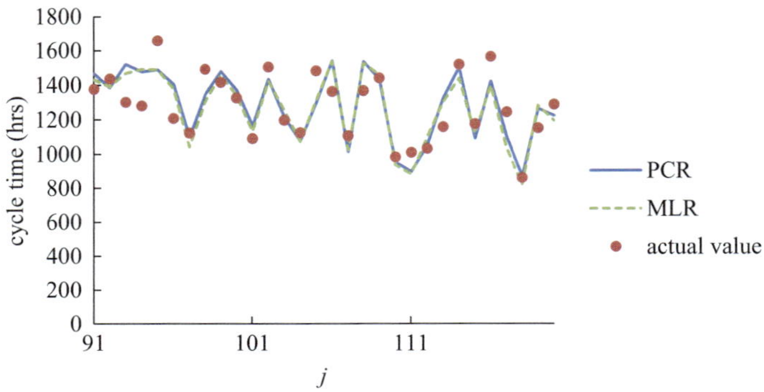

Fig. 2.5 Comparison of the forecasting results using PCR and MLR

$$\text{MAPE} = 7.6\%$$
$$\text{RMSE} = 118 \text{ (hrs)}$$
$$R^2 = 0.61$$

The forecasting results using PCR and MLR are compared in Fig. 2.5. Obviously, the forecasting accuracy using PCR is slightly higher than that using MLR in this case.

2.3 Production Simulation for Cycle Time Prediction

The procedure for conducting production simulation to predict the cycle time of each job in a manufacturing system comprises eight steps [6]:

Step 1. Develop a simulation model for the next scenario: A numerical simulation model that removes visual features such as factory layout and animation is sufficient and more efficient (see Fig. 2.6).

Step 2. Execute the simulation model to simulate the production in the factory during the planning horizon.

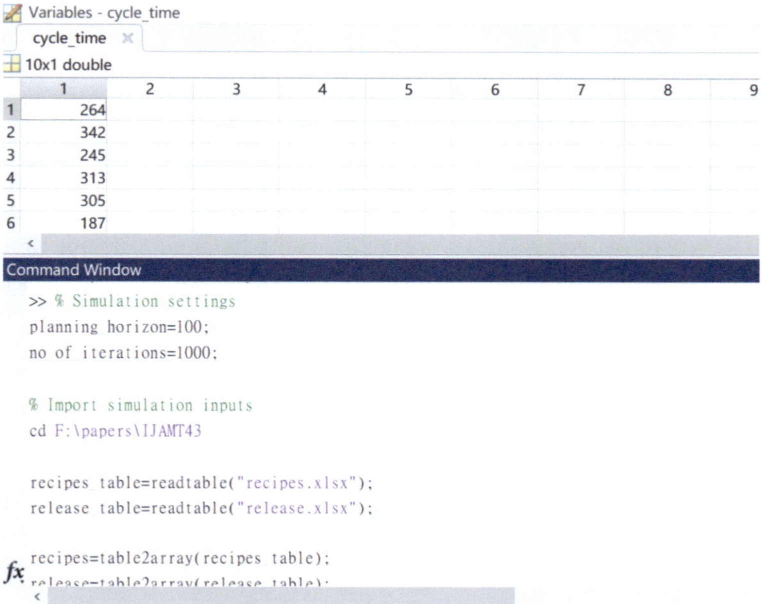

Fig. 2.6 Numerical simulation for cycle time prediction

Step 3. Collect performance measures and generate reports at the end of each simulation run.
Step 4. Repeat Steps 2~3.
Step 5. Generate the statistics of performance measures, such as the mean and variance of cycle times across all simulation runs.
Step 6. If all scenarios have been analyzed, stop; otherwise, get back to Step 1.
Step 7. Compare cycle time means and standard deviations in all scenarios.
Step 8. Fit the relationship between cycle time and any input variable of interest.

In the literature, there have been many studies on the application of production simulation to cycle time prediction. For example, Sivakumar and Chong [7] established a data-driven discrete event simulation model for a semiconductor backend manufacturing system to understand the relationship between controllable input variables and cycle time, thereby reducing the mean and variance of cycle times and increasing throughput. The controllable input variables considered in their study included lot release control, machine scheduling heuristics, the elimination of selected processes, material handling time, setup time, and machine availability.

Many studies have shown that reducing lot size can effectively shorten cycle time. To this end, Wang and Wang [8] developed a simulation model for a wafer fabrication plant (wafer fab) to find the optimal lot size that minimized cycle times under different bottleneck loads.

Vasili et al. [9] derived several analytical equations to estimate the average cycle time of processing jobs in a split-type automated storage/retrieval system (AS/RS) to shorten the average cycle time. The estimation accuracy of these analytical equations was verified by performing Monte Carlo simulation (i.e., discrete event simulation) on the split-type AS/RS.

Yang et al. [10] adopted the generalized gamma distribution to model the distribution of cycle times. Then, they conducted production simulation to fit the cycle time-throughput percentile curve, in which the first three cycle time-throughput moment curves within the range of interest of throughput were fitted by obtaining the parameters of the generalized gamma distribution through matching moments. Any percentile of interest could then be obtained by inverting the distribution.

Ankenman et al. [11] integrated discrete event simulation, adaptive statistical methods, and analytical queue analysis to establish a mathematical equation that quantified the relationship between cycle time and throughput. They considered two scenarios: single product and multiple products, and assumed that the manufacturing system was in a steady state.

Atan et al. [12] used ProModel to simulate the existing production line of a clothing manufacturer to develop and implement improvement measures to shorten cycle time and increase output. Currently, the manufacturer had to work overtime to complete some orders, otherwise it would miss their due dates or lose the orders.

In sum,

- Since most inputs have distributions, a large number of runs need to be simulated, making production simulation a time-consuming process. This phenomenon is more pronounced when simulating large and complex manufacturing systems.
- Since it is impossible to know what the final result will be from these simulation runs, production simulation is often used to compare average performances (i.e., $\overline{C}_j$ rather than C_j) under multiple scenarios.
- Production simulation can be applied to predict the average, standard deviation, and distribution of the cycle time of a job by considering the distributions of all inputs. As a result, the range of the cycle time will be very wide, which makes the actual value highly uncertain.

Nevertheless, production simulation remains one of the most reliable methods for predicting cycle times if the uncertainty in production conditions can be controlled.

2.4 Machine Learning, Deep Learning

2.4.1 Classification and Regression Tree (CART), Random Forest (RF)

Classification and regression trees (CARTs) are a decision tree algorithm that can be used for classification and regression tasks. It is a supervised learning algorithm that predicts unseen data by learning from labeled data [13]. CART builds a tree-like structure consisting of nodes and branches. Nodes represent different decision (or splitting) points and branches represent the possible outcomes of these decisions. The leaf nodes in the tree contain the predicted class labels or values of the target variable.

CART uses a greedy algorithm to split the collected data at each node. It evaluates all possible splits and selects the one that reduces the impurity of the resulting subsets the most. For regression tasks, CART adopts residual reduction (RMSE minimization) as the splitting criterion. In addition, to prevent overfitting, cost complexity pruning is performed, which involves calculating the cost of each node and removing nodes with negative costs. In addition, information gain pruning can also be performed, which involves calculating the information gain of each node and removing nodes with low information gains.

Example 2.4 A CART is built to predict the cycle times of jobs in Example 2.2, for which the Python code is provided in Fig. 2.7. The built CART is illustrated in Fig. 2.8. The forecasting results are summarized in Fig. 2.9. The forecasting accuracy is evaluated as

$$\text{MAE} = 92\,(\text{hrs})$$
$$\text{MAPE} = 7.9\%$$
$$\text{RMSE} = 116\,(\text{hrs})$$
$$R^2 = 0.47$$

CART-based methods have been widely used for cycle time prediction. For example, Wu and Chen [14] proposed a hybrid CART and backpropagation network (BPN) method to predict the cycle time of a job in a wafer fab, in which jobs were first

```
# -*- coding: utf-8 -*-
from sklearn import tree
from sklearn.model_selection import train_test_split
import pandas as pd
import matplotlib.pyplot as plt
cycletimedata = pd.read_excel('F:\\XAI-class\\cycle time forecasting example.xlsx')

# Specify the features you want to use
features = ['xj1', 'xj2', 'xj3', 'xj4', 'xj5', 'xj6']
X = cycletimedata.iloc[0:200, 1:7][features]
y = cycletimedata.iloc[0:200, 7]
X_train, X_test, y_train, y_test = train_test_split(X, y, test_size=0.25, random_state=42)
clf = tree.DecisionTreeRegressor(max_depth=3,
                                 min_samples_split=20,
                                 min_samples_leaf=10)
clf = clf.fit(X, y)

plt.figure(figsize=(20, 10))
tree.plot_tree(clf,
               feature_names=features,
               filled=True,
               rounded=True,
               fontsize=10)
plt.show()
```

Fig. 2.7 Python code for building the CART for Example 2.2

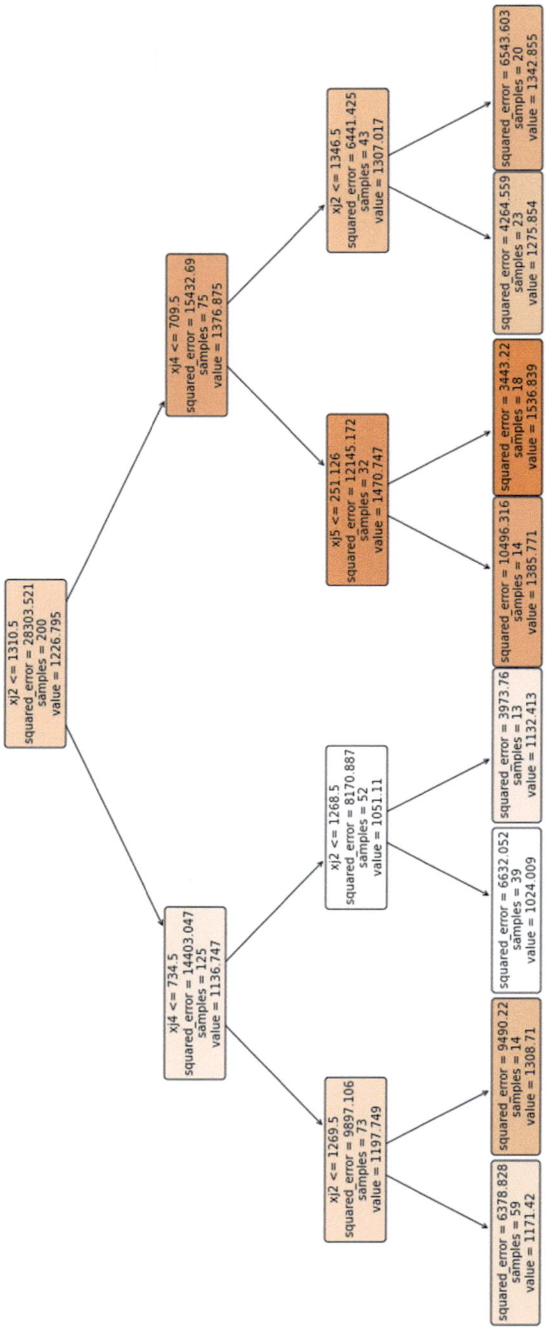

Fig. 2.8 CART built

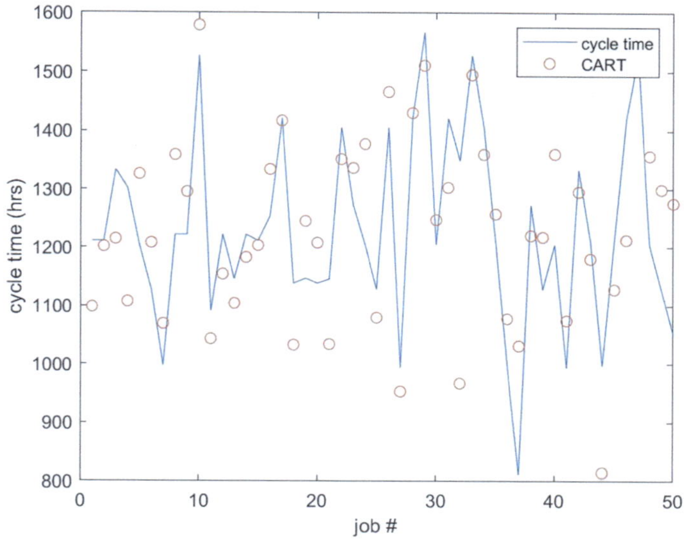

Fig. 2.9 Forecasting results using CART

classified by a CART. Then, a BPN was constructed for each branch of the CART to estimate the cycle times of jobs belonging to this branch. In addition, CART is receiving much attention in the trend of XAI because a CART involves decision rules that are easy to comprehend and communicate. However, a single CART alone may not be sufficiently effective, so methods based on the collaboration of multiple CARTs, such as random forest (RF) [15] and extreme gradient boosting (XGBoost) [16], have been proposed one after another.

An RF can also be constructed to predict the cycle times of jobs. The RF contains B decision trees. Each decision tree considers a random subset of job attributes (i.e., feature bagging) and is trained using examples randomly selected from the data with replacement (i.e., bootstrapping) [17]. In each decision tree, there is a decision rule applicable to a certain job (see Fig. 2.10), which has the following format [18]:

$$\text{“If } x_{j1} \wedge_{q1} (b) c_{q1}(b) \text{ and } \ldots \text{and } x_{jM} \wedge_{qM} (b) \\ c_{qM}(b) \text{ then } \hat{C}_j(b) = d_q(b)\text{”} \tag{2.19}$$

where $c_{qm}(b)$ is the threshold of x_{jm} in the q-th decision rule of decision tree b; $q = 1 \sim Q(b)$ (the number of rules); $b = 1 \sim B$. $\wedge_{qm}(b) \in \{=, <, \geq, within\}$ is the used logical operator; and $d_q(b)$ is the outcome of decision rule q in decision tree b. Each decision tree is trained to minimize the RMSE of approximating $\hat{C}_j(b)$ with C_j, subject to constraints such as the minimum number of jobs of a branch, the maximum number of nodes, and the tallest tree height/depth [17]. The optimization process goes through three phases: tree growth, stopping, and pruning. A number of

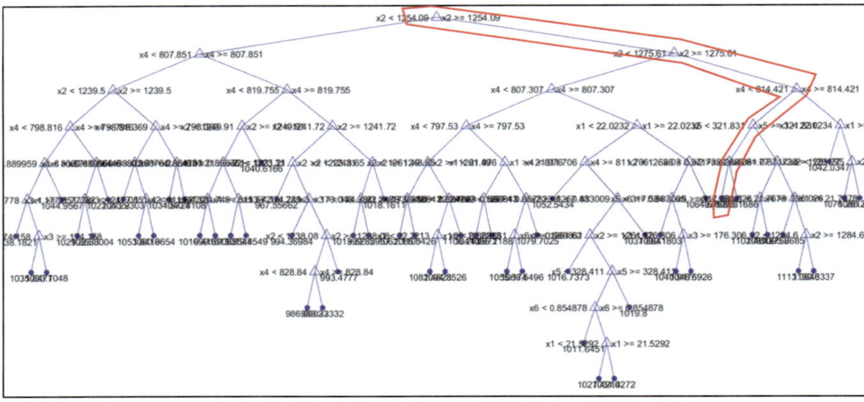

Fig. 2.10 Decision rule applicable to a certain job in each decision tree

decision trees then form the RF after a series of feature bagging and bootstrapping [19].

Example 2.5 The RF algorithm is applied to the collected data in Example 2.2. The required MATLAB code is shown in Fig. 2.11, in which the maximum number of trees is set to 10. The forecasting results are summarized in Fig. 2.12. The 10 trees of the RF are shown in Fig. 2.13.

```
cd F:\ACTM
example2=readtable("example2.xlsx");
example2=table2array(example2);
train_x=example2(1:150, 2:7);
test_x=example2(151:200, 2:7);
train_y=example2(1:150, 8:8);
test_y=example2(151:200, 8:8);

RF1=TreeBagger(10, train_x, train_y);
test_o=predict(RF1, test_x);
test_o=str2double(test_o);
test_rmse=mean((test_y-test_o).^2)^0.5;
test_mae=mean(abs(test_y-test_o));
test_mape=mean(abs(test_y-test_o)./test_y);
test_R2=1-sum((test_y-test_o).^2)/(std(test_y)^2*(50-1));
```

Fig. 2.11 MATLAB code for implementing the RF algorithm

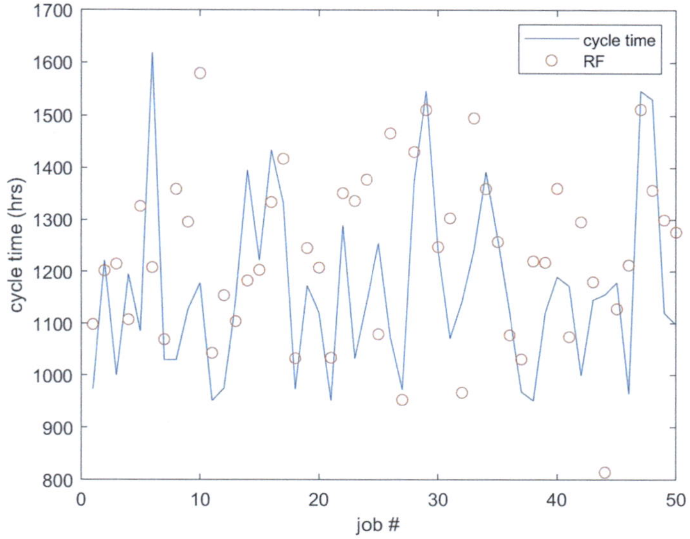

Fig. 2.12 Forecasting results using the RF algorithm

2.4.2 Support Vector Regression (SVR)

Support vector machines (SVMs) are a set of supervised learning methods for classification, regression, and outlier detection [20]. SVMs are very effective for high-dimensional problems. In particular, SVMs are effective when the number of dimensions is larger than the number of examples. SVMs also save memory by using a subset of training points (called support vectors) in the decision function. In addition, SVMs achieve high versatility by specifying different kernel functions for the decision function. SVM can be extended to solve regression problems, usually called support vector regression (SVR).

SVR can be applied to predict the cycle times of jobs as follows [21]:

$$\text{Min}\quad Z = \frac{1}{2}w^T w + K\sum_{j=1}^{n}\left(\zeta_j + \zeta_j^*\right) \tag{2.20}$$

subject to

$$C_j - w^T\phi\left(\mathbf{x_j}\right) - b \leq \varepsilon + \zeta_j \tag{2.21}$$

$$w^T\phi\left(\mathbf{x_j}\right) + b - C_j \leq \varepsilon + \zeta_j^* \tag{2.22}$$

$$\zeta_j, \zeta_j^* \geq 0, j = 1 \sim n \tag{2.23}$$

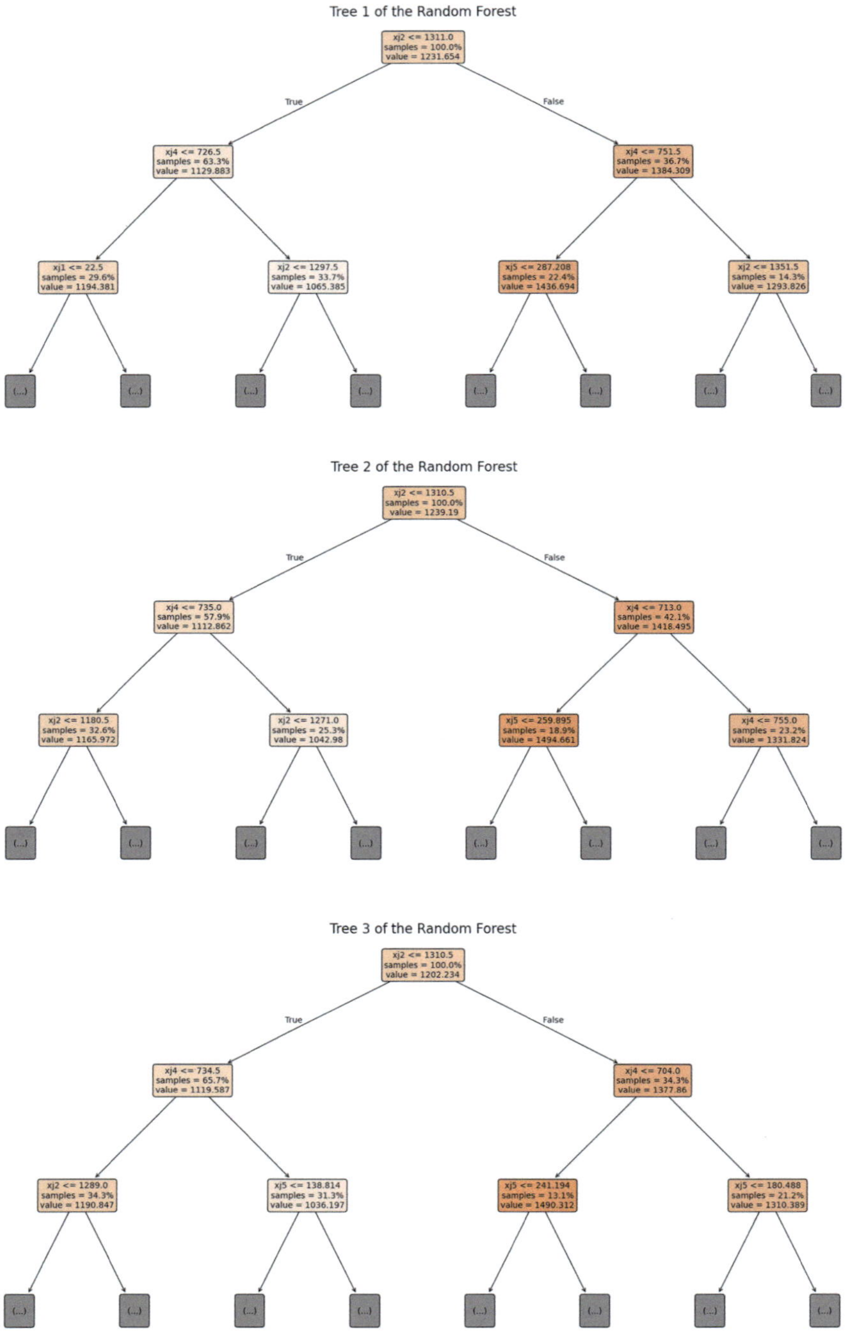

Fig. 2.13 Ten trees of the RF

Table 2.4 Cycle time-related data of 200 jobs

j	x_{j1}	x_{j2}	x_{j3}	x_{j4}	x_{j5}	x_{j6} (%)	C_j (hrs)
1	22	1320	165	748	421	85	1298
2	23	1261	169	734	198	81	1098
3	22	1187	159	635	287	93	1122
…							
200	24	1309	175	734	345	81	1276

$\phi(\mathbf{x_j})^T\phi(\mathbf{x_j})$ is the kernel. The training vectors are implicitly mapped to a higher (possibly infinite) dimensional space via the $\phi()$ function. The cycle time of job j is predicted as

$$\hat{C}_j = w^T\phi(\mathbf{x_j}) + b \tag{2.24}$$

Example 2.6 Cycle time-related data have been collected for 200 jobs (see Table 2.4). SVR is applied to predict the cycle times of jobs in the training data (the first 150 job). The MATLAB program used to implement SVR is shown in Fig. 2.14. The prediction accuracy is evaluated as follows:

$$\begin{aligned} \text{MAE} &= 58 \text{ (hrs)} \\ \text{MAPE} &= 5.0\% \\ \text{RMSE} &= 76 \text{ (hrs)} \\ R^2 &= 0.78 \end{aligned}$$

The cycle time forecasting results are summarized in Fig. 2.15.

2.4.3 Gradient Boosting (GB)

Gradient boosting (GB) is an ensemble learning method for classification and regression tasks. It is a boosting algorithm that combines multiple weak learners to create a strong predictive model. It works by training weak learners sequentially, with each new weak learner trying to correct the mistake made by its predecessor. GB consists of three steps:

- Sequential learning process.
- Residual calculation.
- Shrinkage.

Once all weak learners have been trained, the contributions of all weak learners can be summed up to make a prediction. The cycle time of job j is predicted using GB as

```
% Load the data
cd F:\ACTM
example2=readtable("example2.xlsx");
example2=table2array(example2);

% Split the data
train_x=example2(1:150, 2:7);
test_x=example2(151:200, 2:7);
train_y=example2(1:150, 8:8);
test_y=example2(151:200, 8:8);

% Apply SVR
svm1=fitrsvm(train_x, train_y)

% Generate cycle time forecasts
train_o=predict(svm1, train_x);
test_o=predict(svm1, test_x);

% Evaluate the forecasting accuracy
test_rmse=mean((test_y-test_o).^2)^0.5;
test_mae=mean(abs(test_y-test_o));
test_mape=mean(abs(test_y-test_o)./test_y);
test_R2=1-sum((test_y-test_o).^2)/(std(test_y)^2*(50-1));

% Show the cycle time forecasting results
plot(example2(151:200, 1:1), test_y);
hold on;
plot (example2(151:200, 1:1), test_o, '+');
hold off;
legend({'Cycle Time', 'SVR'});
xlabel('job #');
ylabel('Cycle Time (hrs)');
```

Fig. 2.14 MATLAB for implementing the GB for cycle time prediction

$$\hat{C}_j = \hat{C}_j(1) + \sum_{s=1}^{S} \eta r_j(s) \tag{2.25}$$

where $\hat{C}_j(1)$ is the prediction by the first weak learner; $r_j(s)$ indicates the residual (i.e., prediction error) for job j by weak learner s; and $s = 1 \sim S$ (the number of weak learners).

Example 2.7 GB is applied to predict the cycle times of jobs in Example 2.6. The MATLAB program used to implement the RBF is shown in Fig. 2.16. The aggregation algorithm is least squares boosting (LSBoost). Weak learners are regression trees that allow a maximum of 10 splits. 100 weak learners are used. Therefore, $S = 100$. The prediction accuracy is evaluated as follows:

$$\text{MAE} = 73 \text{ (hrs)}$$
$$\text{MAPE} = 6.2\%$$
$$\text{RMSE} = 100 \text{ (hrs)}$$

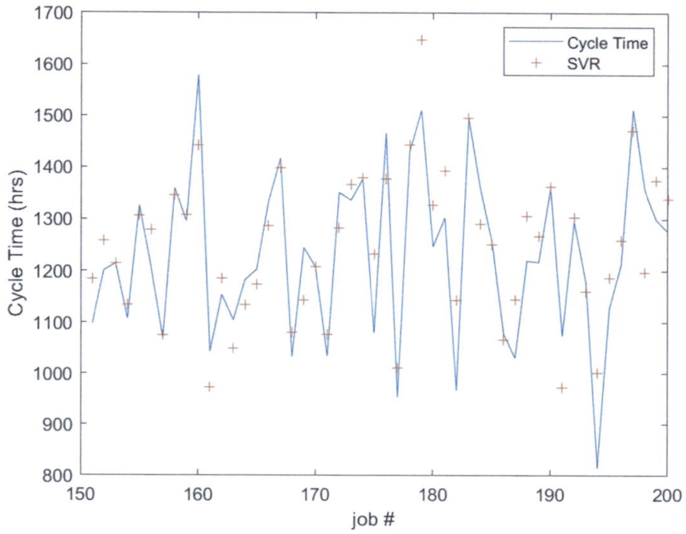

Fig. 2.15 Cycle time forecasting results using SVR

$$R^2 = 0.61$$

The cycle time forecasting results are summarized in Fig. 2.17.

2.4.4 Radial Basis Function Network (RBF), Feedforward Neural Network (FNN), DNN

The first type of ANNs used for cycle time forecasting is radial basis function networks (RBFs) [22]. An RBF network is a special type of feedforward neural network (FNN) that uses a radial basis (or Gaussian) function:

$$h_{jp} = e^{\frac{\left\|\sum_{m=1}^{M}\left(w_{mp}x_{jm}\right)-c_p\right\|^2}{2\sigma_p^2}} \tag{2.26}$$

rather than a logistic sigmoid (logsig) function

$$h_{jp} = \frac{1}{1 + e^{-\left(\sum_{m=1}^{M}\left(w_{mp}x_{jm}\right)-\vartheta_p\right)}} \tag{2.27}$$

as the activation function of hidden layers. In addition, the activation function of the output layer is a linear function (i.e., the weighted sum):

```
% Load the data
cd F:\ACTM
example2=readtable("example2.xlsx");
example2=table2array(example2);

% Split the data
train_x=example2(1:150, 2:7);
test_x=example2(151:200, 2:7);
train_y=example2(1:150, 8:8);
test_y=example2(151:200, 8:8);

% Apply GB
gb1=fitrensemble(train_x, train_y)

% Generate cycle time forecasts
train_o=predict(gb1, train_x);
test_o=predict(gb1, test_x);

% Evaluate the forecasting accuracy
test_rmse=mean((test_y-test_o).^2)^0.5;
test_mae=mean(abs(test_y-test_o));
test_mape=mean(abs(test_y-test_o)./test_y);
test_R2=1-sum((test_y-test_o).^2)/(std(test_y)^2*(50-1));

% Show the cycle time forecasting results
plot(example2(151:200, 1:1), test_y);
hold on;
plot (example2(151:200, 1:1), test_o, '+');
hold off;
legend({'Cycle Time', 'GB'});
xlabel('job #');
ylabel('Cycle Time (hrs)');
```

Fig. 2.16 MATLAB code for applying GB

$$\hat{C}_j = \sum_{p=1}^{P} \left(w_p x_{jp} \right) \tag{2.28}$$

RBFs are widely considered to be one of the most effective types of FNNs [23]. For example, Chen [24] proposed a fuzzy linear regression (FLR)-partial consensus fuzzy intersection (PCFI)-RBF method for accurately predicting future work-in-process (WIP) levels. In the FLR-PCFI-RBF method, a virtual expert committee was formed. Each virtual expert fits an FLR equation to predict the WIP level by solving two nonlinear programming (NLP) problems from its perspective. A two-step aggregation mechanism was adopted to aggregate the fuzzy WIP level forecasts by all virtual experts. First, PCFI was applied to aggregate the fuzzy predictions into a polygonal fuzzy number to improve the precision in terms of the average range of fuzzy WIP level forecasts. Subsequently, an RBF is constructed to defuzzify the polygonal fuzzy number to generate a representative/clear value, thereby improving the accuracy.

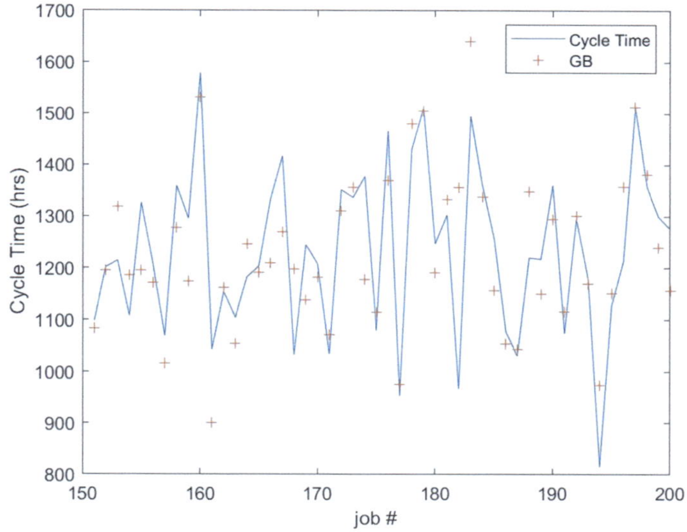

Fig. 2.17 Cycle time forecasting results using GB

Example 2.8 An RBF is constructed to predict the cycle times of jobs in Example 2.6. The MATLAB program used to implement the RBF is shown in Fig. 2.18. The prediction accuracy is evaluated as follows:

$$\text{MAE} = 99\ (\text{hrs})$$
$$\text{MAPE} = 8.1\%$$
$$\text{RMSE} = 125\ (\text{hrs})$$
$$R^2 = 0.39$$

The cycle time forecasting results are summarized in Fig. 2.19.

Wang et al. [22] proposed a cycle time prediction method based on parallel processing, which decomposed the collected cycle time-related data into multiple parts, each of which was then learned by an RBF. The network parameter values of different RBFs were not equal. To select their representative values, the most common value (or average value) of each network parameter was derived. The experimental results showed that the RBF method based on parallel processing outperformed a single RBF, backpropagation network (BPN), and MLR.

The second type of ANNs used for cycle time prediction is a multilayer perceptron (or FNN) [15]. The architecture of an FNN is shown in Fig. 2.20, which has $2 + Q$ layers: an input layer, Q hidden layers, and an output layer. When an FNN has more than one hidden layer, it becomes a DNN.

In the network training phase, inputs to the FNN are weighted and transmitted to each node of the first hidden layer, on which they are aggregated as

```
% Load the data
cd F:\ACTM
example2=readtable("example2.xlsx");
example2=table2array(example2);

% Split the data
train_x=example2(1:150, 2:7);
test_x=example2(151:200, 2:7);
train_y=example2(1:150, 8:8);
test_y=example2(151:200, 8:8);

% newrb(input, output, MSE target, spread, maximum number of hidden-layer nodes, increment of
the number of hidden-layer nodes)
net1=newrb(train_x', train_y', 0, 30, 50, 1);

% Generate cycle time forecasts
train_o=sim(net1, train_x');
test_o=sim(net1, test_x');

% Evaluate the forecasting accuracy
test_rmse=mean((test_y-test_o').^2)^0.5;
test_mae=mean(abs(test_y-test_o'));
test_mape=mean(abs(test_y-test_o')./test_y);
test_R2=1-sum((test_y-test_o').^2)/(std(test_y)^2*(50-1));

% Show the cycle time forecasting results
plot(example2(151:200, 1:1), test_y);
hold on;
plot (example2(151:200, 1:1), test_o, '+');
hold off;
legend({'Cycle Time', 'RBF'});
xlabel('job #');
ylabel('Cycle Time (hrs)');
```

Fig. 2.18 MATLAB code for implementing the RBF for cycle time forecasting

$$I_{jl_1}^{h(1)} = \sum_{p=1}^{P} \left(w_{pl_1}^{h(1)} x_{jp} \right) \tag{2.29}$$

where x_{jp} is the p-th input of job j; $p = 1 \sim P$; $j = 1 \sim n$. After subtracting the threshold on the hidden-layer node, the output from the node is derived as

$$h_{jl_1}^{(1)} = \frac{2}{1 + e^{-2n_{jl_1}^{h(1)}}} - 1 \tag{2.30}$$

in which

$$n_{jl_1}^{h(1)} = I_{jl_1}^{h(1)} - \vartheta_{l_1}^{h(1)} \tag{2.31}$$

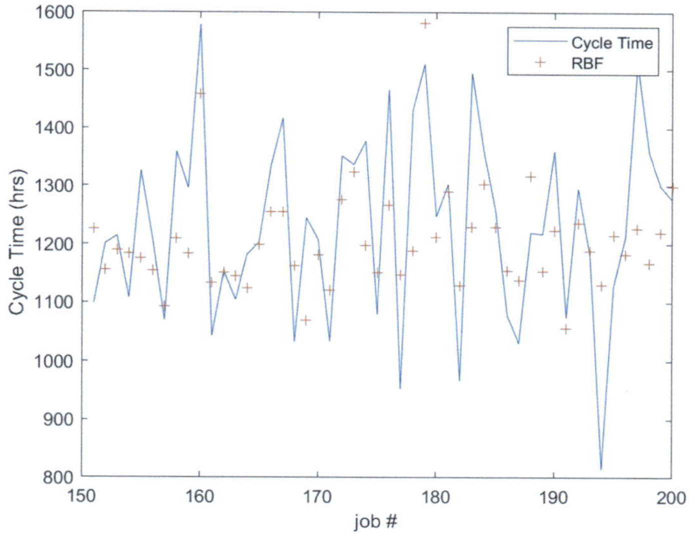

Fig. 2.19 Cycle time forecasting results using RBF

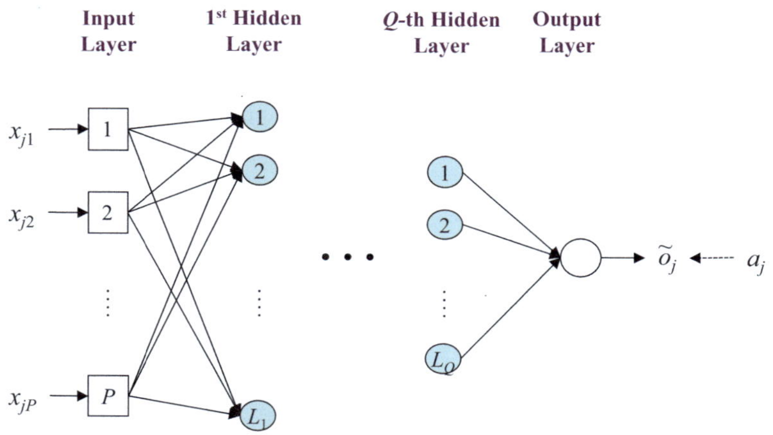

Fig. 2.20 Architecture of FNN for cycle time forecasting

where $h_{jl_1}^{(1)}$ is the output from node l_1 of the first hidden layer; $l_1 = 1 \sim L_1$. $\vartheta_{l_1}^{h(1)}$ is the threshold on this node; and $w_{pl_1}^{h(1)}$ is the connection weight between input node p and this node. It is worth noting that hyperbolic tangent (tanh) activation function is used in Eq. (2.21) rather than sigmoid function:

$$h_{jl_1}^{(1)} = \frac{1}{1 + e^{-n_{jl_1}^{h(1)}}} \tag{2.32}$$

where $h_{jl_1}^{(1)}$ is passed to the second hidden layer and processed in the same way. Finally, outputs from the last hidden layer are weighted and transmitted to the output node, on which they are aggregated as

$$I_j^o = \sum_{p=1}^{P} \left(w_{l_Q}^o h_{jl_Q}^{(Q)} \right) \tag{2.33}$$

where $w_{l_Q}^o$ is the connection weight between node l_Q of the Q-th hidden layer and the output node; $h_{jl_Q}^{(Q)}$ is the output from this node for job j. Then, the network output $\hat{C}_j$ is generated as

$$\hat{C}_j = n_j^o \tag{2.34}$$

where

$$n_j^o = I_j^o - \vartheta^o \tag{2.35}$$

in which the linear function is employed. The most efficient algorithms for training the FNN include the Levenberg–Marquardt (LM) algorithm [25].

Example 2.9 An FNN with three hidden layers is constructed to predict the cycle times of jobs in Example 2.8. The numbers of nodes in the hidden layers are 12, 6, and 6, respectively. The MATLAB program used to construct the FNN is shown in Fig. 2.21. The prediction accuracy is evaluated as follows:

$$\begin{aligned} \text{MAE} &= 73\ \text{(hrs)} \\ \text{MAPE} &= 6.0\% \\ \text{RMSE} &= 103\ \text{(hrs)} \\ R^2 &= 0.58 \end{aligned}$$

The cycle time forecasting results are summarized in Fig. 2.22.

There are very few cases where adding a second (or third, etc.) hidden layer will improve cycle time prediction performance. For most problems, one hidden layer is sufficient, as shown in the example below.

Example 2.10 A parametric analysis was performed to examine the effect of the number of hidden layers (see Fig. 2.23) on the prediction accuracy. The cycle time forecasting results are summarized in Fig. 2.24. The prediction performances achieved using different numbers of hidden layers are compared as follows:

```
% Load the data
cd F:\ACTM
example2=readtable("example2.xlsx");
example2=table2array(example2);

% Split the data
train_x=example2(1:150, 2:7);
test_x=example2(151:200, 2:7);
train_y=example2(1:150, 8:8);
test_y=example2(151:200, 8:8);
train_x=transpose(train_x);
train_y=transpose(train_y);
test_x=transpose(test_x);
test_y=transpose(test_y);

% Configure the FNN
net1=feedforwardnet([12, 6, 6]);
net1.dividefcn='dividetrain'; % allocate all samples to the training data
net1.trainParam.lr=0.1;
net1.trainParam.epochs=20000;
net1.trainParam.goal=100;

% Train the FNN
net1=train(net1, train_x, train_y);

% Generate cycle time forecasts
train_o=net1(train_x);
test_o=net1(test_x);

% Evaluate the forecasting accuracy
test_rmse=mean((test_y-test_o).^2)^0.5;
test_mae=mean(abs(test_y-test_o));
test_mape=mean(abs(test_y-test_o)./test_y);
test_R2=1-sum((test_y-test_o).^2)/(std(test_y)^2*(50-1));

% Show the cycle time forecasting results
plot(example2(151:200, 1:1), test_y);
hold on;
plot (example2(151:200, 1:1), test_o, '+');
hold off;
legend({'Cycle Time', 'FNN'});
xlabel('job #');
ylabel('Cycle Time (hrs)');
```

Fig. 2.21 MATLAB code for constructed FNN for cycle time forecasting

$$\text{RMSE (1 hidden layer)} = 51 \text{ (hrs)}$$
$$\text{RMSE (2 hidden layers)} = 79 \text{ (hrs)}$$
$$\text{RMSE (3 hidden layers)} = 90 \text{ (hrs)}$$

In this example, when the FNN has a single layer, it achieved the highest cycle time prediction accuracy in terms of RMSE.

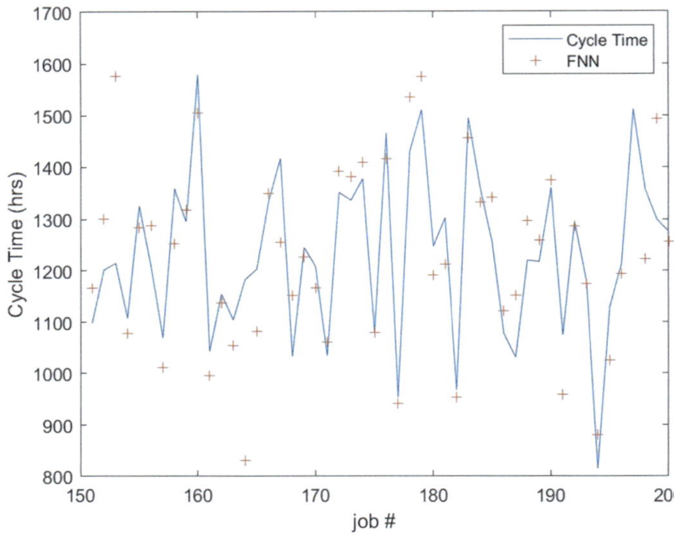

Fig. 2.22 Cycle time forecasting results using FNN

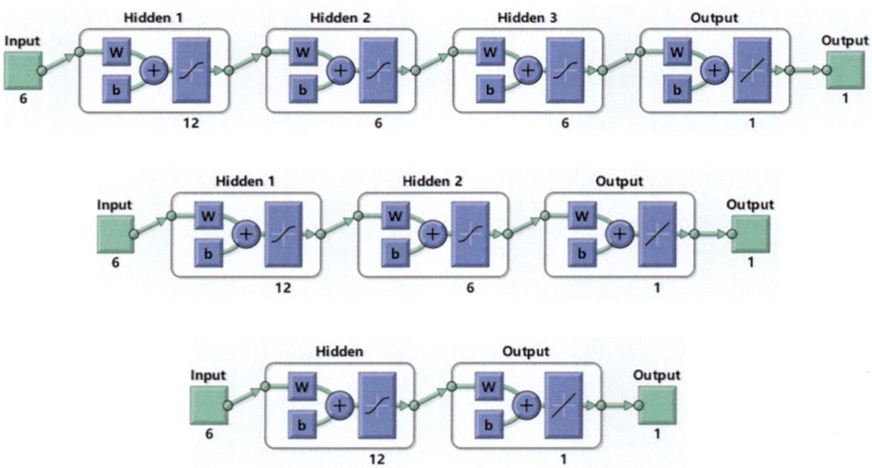

Fig. 2.23 FNN with various numbers of hidden layers

2.4.5 *Self-organization Map (SOM)-FNN*

Classifying jobs before predicting their cycle times has proven to be an effective treatment [14, 26, 27]. Chen [27] proposed a hybrid self-organization map (SOM)-FNN method to predict the completion times of jobs in a wafer fab, which is equivalent to predicting their cycle times since the release times were fixed.

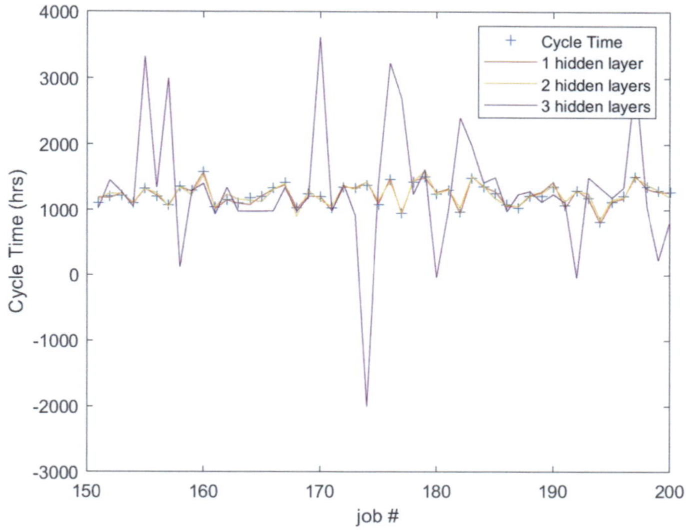

Fig. 2.24 Cycle time forecasting results with various numbers of hidden layers

Each job inputted to the FNN is called an example. Before being inputted to the FNN, the examples are pre-classified into different categories by an SOM. The structure of the SOM is shown in Fig. 2.25. It is a 10×10 SOM, that is, the number of output nodes is 100.

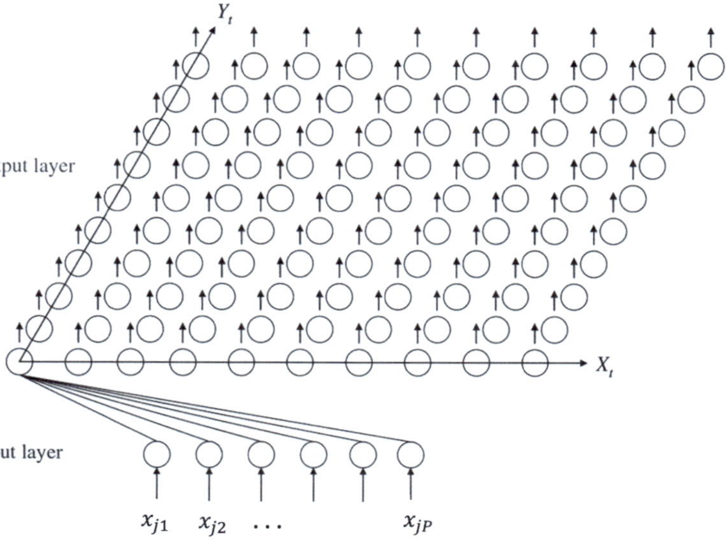

Fig. 2.25 Architecture of the SOM

The feature vectors of all jobs are input into the SOM for learning as follows:

Step 1. Set the number of output nodes and the number of input nodes. Initialize the learning rate, neighborhood size, and number of iterations.
Step 2. Initialize the weights (w_{lp}) randomly, where $l = 1 \sim L$ and L stands for the number of output nodes; $p = 1 \sim P$.
Step 3. (Iteration) Provide an input vector to the network.
Step 4. Find the output node (winner) based on the similarity between the input vector and the weight vector. For an input vector $\mathbf{x_j}$, the winning unit can be determined by the distance $||\mathbf{x_j} - \mathbf{w_c}|| = \min_l ||\mathbf{x_j} - \mathbf{w_l}||$, where $\mathbf{w_l}$ is the weight vector of the l-th unit and index c refers to the winning unit.
Step 5. Update the weight vector of the winner node using Kohonen's learning rule [28]:

$$\mathbf{w_l}(t+1) = \mathbf{w_l}(t) + \alpha(t)\left(\mathbf{x_j} - \mathbf{w_l}\right) \text{ for each } l \in N_c(t) \tag{2.36}$$

where t is the time index; $\alpha(t) \in [0, 1]$ is a scalar that defines the relative size of the learning step; and $N_c(t)$ specifies the neighborhood around the winner in the map array.
Step 6. Stop if the number of iterations has been run. Otherwise, go to Step 3.

Example 2.11 To enhance the accuracy of cycle time prediction, jobs in Example 2.10 are classified using a 5 × 2 SOM that classifies jobs into at most 10 categories. The MATLAB code for this purpose is provided in Fig. 2.26. The trained SOM is illustrated in Fig. 2.27. In this figure, blue hexagons represent neurons. Red lines connect neighboring neurons. Regions containing red lines indicate the distances between neurons. Darker areas represent larger distances, while lighter areas represent smaller distances. The classification results for the training data are summarized in Fig. 2.28. For example, 19 jobs (job #8, #21, #31, #63, #64, etc.) belong to the 10-th job category.

Example 2.12 Taking the 10th job category as an example, the cycle times of jobs in this category are predicted using an FNN with and without classification for comparison. The FNN is trained using the 19 jobs belonging to this category in the training data and then applied to the 7 jobs belonging to this category in test data. The MATLAB code is provided in Fig. 2.29. The comparison of the prediction results is shown in Fig. 2.30.

```
% Load the data
cd F:\ACTM
example2=readtable("example2.xlsx");
example2=table2array(example2);

% Split the data
train_x=example2(1:150, 2:7);
test_x=example2(151:200, 2:7);
train_y=example2(1:150, 8:8);
test_y=example2(151:200, 8:8);

% Create the SOM
dimension1=5;
dimension2=2;
som1=selforgmap([dimension1 dimension2]);

% Train the SOM
[som1, tr]=train(som1, train_x');

% Classify the examples
train_c=som1(train_x');
test_c=som1(test_x');

% View the SOM
view(som1);

% Identify the category of each example in the training data
category1=zeros(150, 1);
for j=1:150
   for l=1:10
      if train_c(l, j)==1
         category1(j, 1)=l;
      end
   end
end

% Identify the category of each example in test data
category2=zeros(50, 1);
for j=1:50
   for l=1:10
      if test_c(l, j)==1
         category2(j, 1)=l;
      end
   end
end

% Show the classification results
plot(example2(1:150, 1:1), category1);
title('Job Classification Results Using SOM');
xlabel('job #');
ylabel('Category');
```

Fig. 2.26 MATLAB code for the implementation of an SOM

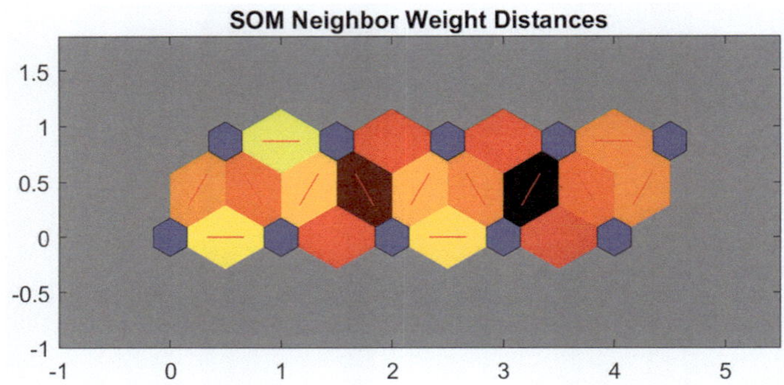

Fig. 2.27 Trained SOM

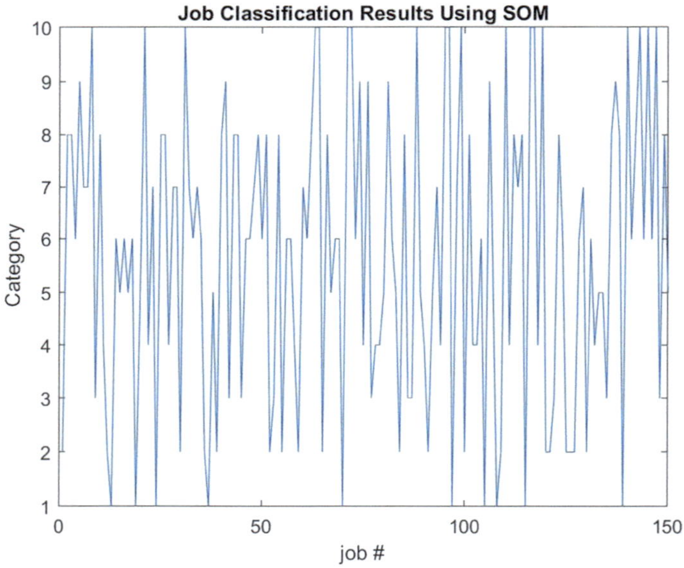

Fig. 2.28 Classification results using SOM (the training data)

2.5 Industry 4.0 Applications in Cycle Time Prediction

2.5.1 Industry 4.0

Industry 4.0 is expected to improve the functionality of manufacturing systems, including cycle time management [29]. Industry 4.0 is equipped with many advanced IT tools, such as cyber-physical systems (CPS), digital twins, 5G and the Internet

```
% Find jobs belonging to the 10th category in the training data
n1=sum(category1==10);
train_x1=zeros(6, n1);
train_y1=zeros(1, n1);
train_o1=zeros(1, n1); % Cycle time forecasts without classification
job_no1=zeros(n1, 1);

count1=0;
for j=1:150
   if category1(j, 1)==10
      count1=count1+1;
      job_no1(count1, 1)=j;
      for p=1:6
         train_x1(p, count1)=train_x(j, p);
      end
      train_y1(1, count1)=train_y(j, 1);
      train_o1(1, count1)=train_o(1, j);
   end
end

% Find jobs belonging to the 10th category in test data
n2=sum(category2==10);
test_x1=zeros(6, n2);
test_y1=zeros(1, n2);
test_o1=zeros(1, n2);
job_no2=zeros(n2, 1);

count2=0;
for j=1:50
   if category2(j, 1)==10
      count2=count2+1;
      job_no2(count2)=j;
      for p=1:6
         test_x1(p, count2)=test_x(j, p);
      end
      test_y1(1, count2)=test_y(j, 1);
      test_o1(1, count2)=test_o(1, j);
   end
end

% Train the FNN
net1=feedforwardnet([12]);
net1.dividefcn='dividetrain'; % allocate all samples to the training data
net1.trainParam.lr=0.1;
net1.trainParam.epochs=20000;
net1.trainParam.goal=100;
net1=train(net1, train_x1, train_y1);
train_o2=net1(train_x1);
test_o2=net1(test_x1);

% Show and compare the forecasting results with and without classification
hold on;
plot(job_no2, test_o2);
plot (job_no2, test_o1);
plot (job_no2, test_y1, '+');
hold off;
legend({'Cycle Time Forecast (Classified)', 'Cycle Time Forecast (Not Classified)', 'Actual Value'});
title('Job Category #10')
xlabel('Job #');
ylabel('Cycle Time (hrs)');
```

Fig. 2.29 Cycle time forecasting using FNN after job classification using SOM

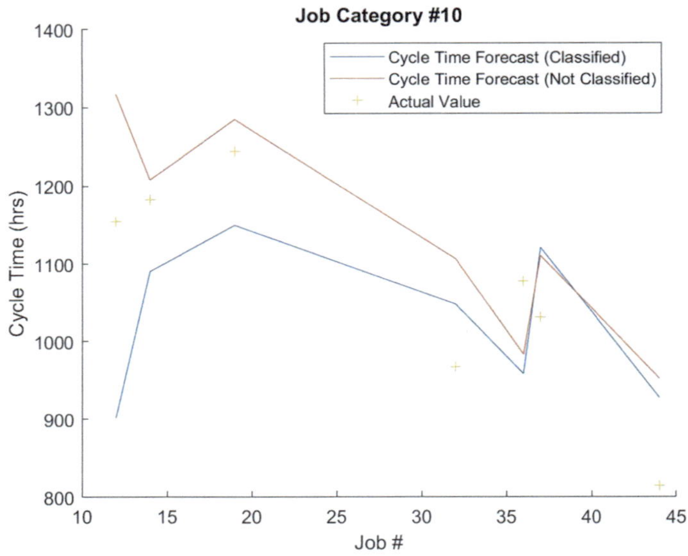

Fig. 2.30 Comparison of the prediction results with and without classification

of things (IoT), massive data (large collections of records in simple formats), ubiquitous computing, big data analytics, cloud computing, cybersecurity, distributed intelligence, and edge computing [30, 31], as shown in Fig. 2.31.

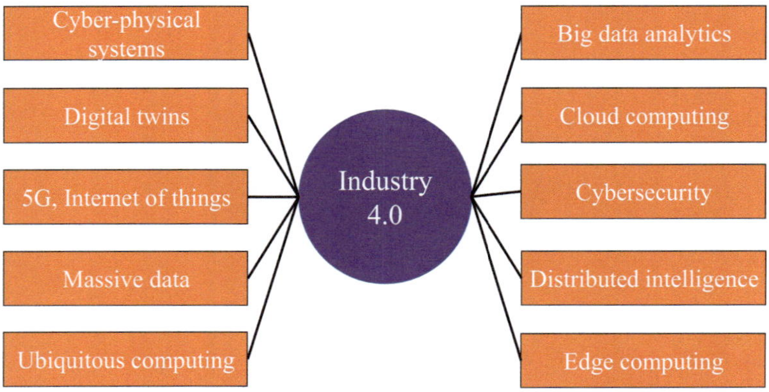

Fig. 2.31 Tools for industry 4.0

2.5.2 Industry 4.0 and Cycle Time Management

Many machines are now controlled by minicomputers to facilitate the fine processing of materials. For example, in a wafer fab, chemical delivery systems, cleaning equipment, chemical vapor deposition (CVD) and physical vapor deposition (PVD) equipment, diffusion furnaces, etc. are embedded with sensors and computers and can therefore be considered CPSs [32]. The state variable identification (SVID) data collected by such sensors can be fed into the DL model for cycle time prediction and other purposes [31].

In addition, radio frequency identification (RFID) tags attached to wafer carriers can be used. Forked light barriers attached to the carrier are controlled by an ATMega32 microcontroller and can detect wafers that have completed manufacturing and are ejected by the equipment [33]. In this way, the start and completion times of each operation on a specific piece of equipment can be immediately and accurately known. Such data is also fed into the DL model for cycle time prediction.

Detecting possible anomalies in wafer manufacturing equipment based on data collected by embedded sensors can reduce quality issues by 35%, increase production capacity by 10%, and increase the yield learning speed by 25%, ultimately helping to shorten cycle times [31]. Such collected data can also be used as inputs to a DL model to predict cycle times.

2.5.3 Big Data Analytics for Cycle Time Prediction

Big data analytics has been considered as the enabler of Industry 4.0 [34]. There are many production conditions related to job cycle time prediction, making it a big data analytics problem. In Industry 4.0, CPSs collect more and more real-time information, which can be used as a reference for predicting the cycle times of jobs, making this big data problem even more difficult. The reference data for predicting the cycle time of a job include job attributes, equipment status, factory conditions, recent production performance, future production plans, etc. [35]. In the literature, stacked sparse autoencoders (SSAE) have been used to address the big data issue in job cycle time prediction [36], thereby constructing an SSAE-FNN to effectively predict the cycle times of jobs.

An autoencoder (AE) is a symmetric neural network whose outputs are designed to reproduce the inputted examples in order to learn the representation of latent features. The basic architecture of an AE is shown in Fig. 2.32. It consists of the input layer, a hidden layer, and the output layer. The encoder part converts job attributes $\{x_{jp}\}$ into latent features $\{h_{jq}\}$:

$$\text{Encoder: } \mathbf{h_j} = \sigma\left(\mathbf{w_1}\mathbf{x_j} + \mathbf{b_x}\right); j = 1 \sim n \tag{2.37}$$

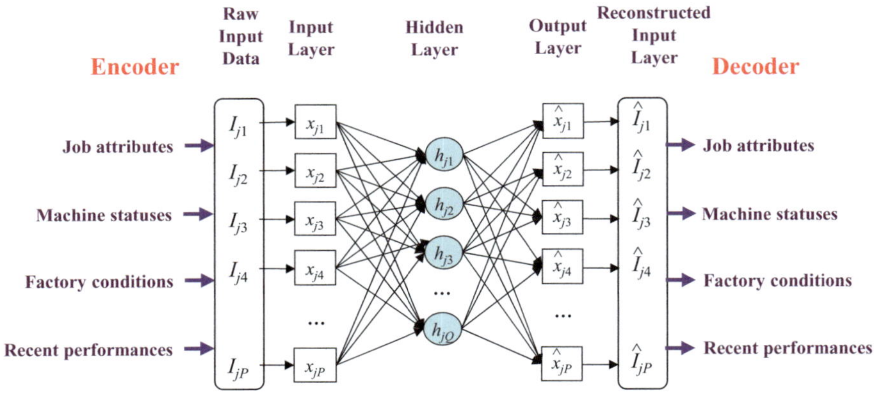

Fig. 2.32 AE for cycle time prediction

In contrast, the output layer is a decoder that reconstructs the latent features $\{h_{jq}\}$ into approximated input values $\{\hat{x}_{jp}\}$:

$$\text{Decoder: } \hat{\mathbf{x}}_{\mathbf{j}} = \sigma'\left(\mathbf{w}_2\mathbf{h}_{\mathbf{j}} + \mathbf{b}_{\mathbf{h}}\right); j = 1 \sim n \tag{2.38}$$

where $\mathbf{w}_1$, $\mathbf{w}_2$, $\mathbf{b}_{\mathbf{x}}$, and $\mathbf{b}_{\mathbf{h}}$ are the weight matrices and biases of the encoder and decoder, respectively. σ and σ' denote the activation functions, using the sigmoid function or the linear rectification function (ReLU). Compared to PCA, which is essentially a linear transformation, autoencoders are able to model complex nonlinear functions. As stated by Tirkel [37], the causal relationship between the cycle time of a job and its attributes is nonlinear, so an autoencoder was more suitable than PCA.

Example 2.13 To predict the cycle time of a job, 35 job attributes and production conditions when the job is released into the factory have been collected (see Table 2.5), including the release time, job size, factory WIP level, the queue lengths before the top ten bottlenecks, route queue length, the average waiting times of recently completed jobs, factory utilization, and the estimated numbers of failures of the top ten bottlenecks, most of which are either retrieved from the production management information system (PROMIS) of the factory or detected by sensors or microcomputers attached to machines.

To reduce the number of inputs to the DL model for cycle time prediction, an AE is constructed to find 10 latent variables from the 35 inputs, for which the MATLAB code is provided in Fig. 2.33. The latent variables for job in the training data are summarized in Table 2.6.

Example 2.14 An FNN is constructed to predict the cycle times of jobs based on their latent variables. The MATLAB code is provided in Fig. 2.34. The forecasting results are summarized in Fig. 2.35.

Table 2.5 Cycle time-related data

j	Release time	Job size	Factory WIP	First bottleneck queue length	…	Route queue length	First job waiting time	…	Factory utilization	First bottleneck est. no. failures	…	Cycle time
1	187	24	1383	184		930	537		0.88	0.96		1419
2	188	22	1340	186		901	518		0.95	1.48		1360
3	198	24	1376	169		911	510		0.83	1.47		1164
	…											
1000	8157	22	1419	174		873	523		0.88	1.24		1383

```
% Load the data
cd F:\ACTM
example_AE=readtable("example_AE.xlsx");
example_AE=table2array(example_AE);

% Split the data
train_x=example_AE(1:800, 2:36);
test_x=example_AE(801:1000, 2:36);
train_y=example_AE(1:800, 37:37);
test_y=example_AE(801:1000, 37:37);

% Build the AE
autoenc1=trainAutoencoder(train_x', 10);

% Convert the training data
train_h=zeros(800, 10);
for j=1:800
    for k=1:35
        for l=1:10
            train_h(j,l)=1/(1+exp(-(autoenc1.EncoderWeights(l,k)*train_x(j,k)+autoenc1.EncoderBiases
    (l,1))));
        end
    end
end

% Convert test data
test_h=zeros(200, 10);
for j=1:200
    for k=1:35
        for l=1:10
            test_h(j, l)=1/(1+exp(-(autoenc1.EncoderWeights(l, k)*test_x(j,k)+autoenc1.EncoderBiases
    (l, 1))));
        end
    end
end
```

Fig. 2.33 MATLAB code for building the AE

Table 2.6 Latent variables for jobs in the training data

j	h_{j1}	h_{j2}	h_{j3}	h_{j4}	h_{j5}	h_{j6}	h_{j7}	h_{j8}	h_{j9}	h_{j10}	Cycle time
1	0.999	0.112	0.443	0.081	0.316	0.428	0.920	0.147	0.565	0.760	1419
2	0.999	0.114	0.429	0.077	0.303	0.429	0.920	0.135	0.556	0.782	1360
3	0.999	0.115	0.418	0.075	0.293	0.430	0.919	0.126	0.550	0.799	1164
	…										
1000	0.999	0.113	0.434	0.078	0.308	0.429	0.920	0.139	0.559	0.774	1383

However, simple autoencoders are unable to effectively extract latent features [36]. Fang et al. [36] therefore proposed the use of sparse autoencoders (SAEs), which is an extension of the basic autoencoder that extracts more meaningful features by adopting a sparse penalty mechanism in the autoencoder. The sparse penalty mechanism is used to control the number of “inactive” and “active” neurons in the hidden layer,

```
% Train the FNN
net1=feedforwardnet([20, 10]);
net1.dividefcn='dividetrain'; % allocate all samples to the training data
net1.trainParam.lr=0.25;
net1.trainParam.epochs=20000;
net1.trainParam.goal=100;
net1=train(net1, train_h', train_y');
train_o=net1(train_h');
test_o=net1(test_h');

% Evaluate the forecasting accuracy
test_rmse=mean((test_y-test_o').^2)^0.5;
test_mae=mean(abs(test_y-test_o'));
test_mape=mean(abs(test_y-test_o')./test_y);
test_R2=1-sum((test_y-test_o').^2)/(std(test_y)^2*(200-1));

% Show the cycle time forecasting results
plot(example_AE(801:1000, 1:1), test_y);
hold on;
plot (example_AE(801:1000, 1:1), test_o, '+');
hold off;
legend({'Cycle Time', 'AE+FNN'});
xlabel('job #');
ylabel('Cycle Time (hrs)');
```

Fig. 2.34 MATLAB code for cycle time prediction based on the latent variables

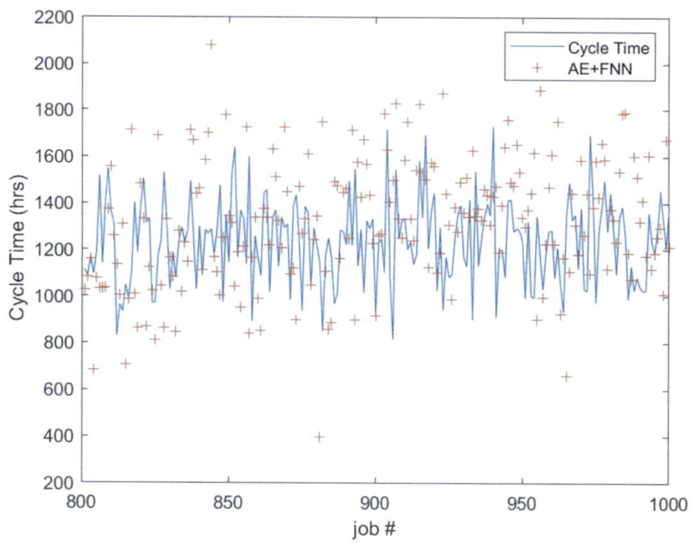

Fig. 2.35 Cycle time forecasting results based on the latent variables

so that the learned features are constrained rather than just copying the input values. The cost function of an SAE is as follows:

$$f(\mathbf{w}_*, \mathbf{b}_*) = \text{MSE} + \gamma \sum_{q=1}^{Q} KL(\rho_q \| \hat{\rho}_q) \tag{2.39}$$

where

$$\text{MSE} = \frac{\sum_{j=1}^{n} \sum_{p=1}^{P} (x_{jp} - \hat{x}_{jp})^2}{n} \tag{2.40}$$

in which $\sum_{q=1}^{Q} KL(\rho_q \| \hat{\rho}_q)$ is the Kullback–Leibler divergence between the average excitation and the expected excitation of hidden-layer neurons. The training process first forward passes all the training data to calculate $\hat{\rho}_q$ and *MSE*, and then uses the backward pass method to optimize the values of $\mathbf{w}_*$ and $\mathbf{b}_*$.

A stacked sparse autoencoder (SSAE) model is constructed by stacking multiple SAEs as hidden layers to construct a DNN. The next stacked layer takes the outputs from the previous stacked layer as inputs. In addition, a dropout layer [38] and a batch normalization layer [39] are also added to the SSAE. First, a dropout layer is added after the SAE and the fully connected layer to avoid overfitting [40, 41]. Next, a batch normalization layer is added after each stacked layer to normalize its output to increase the independence of each layer and accelerate DNN training.

References

1. T. Chen, A fuzzy-neural and multiple-bucket approach for estimating lot cycle time in a wafer fab with dynamic product mix. Comput. Ind. Eng. **55**(2), 423–438 (2008)
2. T.C.T. Chen, Cycle time prediction and output projection, in *Production Planning and Control in Semiconductor Manufacturing: Big Data Analytics and Industry 4.0 Applications* (2022), pp. 41–62
3. I. Mejri, A. Bouzid, S. Bacha, S.B. Layeb, Forecasting demand using ARIMA model and LSTM neural network: a case of detergent manufacturing industry, in *2021 International Conference on Innovation and Intelligence for Informatics, Computing, and Technologies* (2021), pp. 346–353
4. T. Chen, A fuzzy-neural knowledge-based system for job completion time prediction and internal due date assignment in a wafer fabrication plant. Int. J. Syst. Sci. **40**(8), 889–902 (2009)
5. R. Pramoditha, How do you apply PCA to logistic regression to remove multicollinearity? (2021). https://towardsdatascience.com/how-do-you-apply-pca-to-logistic-regression-to-remove-multicollinearity-10b7f8e89f9b
6. C. Harrell, B.K. Ghosh, R.O. Bowden, *Simulation Using ProModel* (McGraw Hill, 2012)
7. A.I. Sivakumar, C.S. Chong, A simulation based analysis of cycle time distribution, and throughput in semiconductor backend manufacturing. Comput. Ind. **45**(1), 59–78 (2001)
8. C.N. Wang, C.H. Wang, A simulated model for cycle time reduction by acquiring optimal lot size in semiconductor manufacturing. The International Journal of Advanced Manufacturing Technology **34**, 1008–1015 (2007)

9. M. Vasili, T.S. Hong, S.M. Homayouni, N. Ismail, A statistical model for expected cycle time of SP-AS/RS: an application of Monte Carlo simulation. Appl. Artif. Intell. **22**(7–8), 824–840 (2008)
10. F. Yang, B.E. Ankenman, B.L. Nelson, Estimating cycle time percentile curves for manufacturing systems via simulation. INFORMS J. Comput. **20**(4), 628–643 (2008)
11. B.E. Ankenman, J.M. Bekki, J. Fowler, G.T. Mackulak, B.L. Nelson, F. Yang, Simulation in production planning: an overview with emphasis on recent developments in cycle time estimation, in *Planning Production and Inventories in the Extended Enterprise: A State of the Art Handbook* vol. 1 (2010), pp. 565–591
12. S.A. Atan, R. Ramlan, T.G. Foong, Cycle time reduction of a garment manufacturing company using simulation technique, in *Proceedings International Conference of Technology Management, Business and Entrepreneurship* (2012), pp. 124–131
13. F.S. Aguiar, L.L. Almeida, A. Ruffino-Netto, A.L. Kritski, F.C. Mello, G.L. Werneck, Classification and regression tree (CART) model to predict pulmonary tuberculosis in hospitalized patients. BMC Pulm. Med. **12**, 1–8 (2012)
14. H.C. Wu, T. Chen, CART–BPN approach for estimating cycle time in wafer fabrication. J. Ambient. Intell. Humaniz. Comput. **6**(1), 57–67 (2015)
15. T. Chen, Y.C. Wang, A modified random forest incremental interpretation method for explaining artificial and deep neural networks in cycle time prediction. Decis. Analytics J. **7**, 100226 (2023)
16. Y. Zhang, X. Shi, S. Zhang, A. Abraham, A xgboost-based lane change prediction on time series data using feature engineering for autopilot vehicles. IEEE Trans. Intell. Transp. Syst. **23**(10), 19187–19200 (2022)
17. V. Rodriguez-Galiano, M. Sanchez-Castillo, M. Chica-Olmo, M.J.O.G.R. Chica-Rivas, Machine learning predictive models for mineral prospectivity: an evaluation of neural networks, random forest, regression trees and support vector machines. Ore Geol. Rev. **71**, 804–818 (2015)
18. J. van der Waa, E. Nieuwburg, A. Cremers, M. Neerincx, Evaluating XAI: a comparison of rule-based and example-based explanations. Artif. Intell. **291**, 103404 (2021)
19. T. Ooka, H. Johno, K. Nakamoto, Y. Yoda, H. Yokomichi, Z. Yamagata, Random forest approach for determining risk prediction and predictive factors of type 2 diabetes: large-scale health check-up data in Japan. BMJ Nutr., Prevention Health **4**(1), 140 (2021)
20. S. Suthaharan, Support vector machine, in *Machine Learning Models and Algorithms for Big Data Classification: Thinking with Examples for Effective Learning* (2016), pp. 207–235
21. Scikit-learn.org, 1.4. *Support Vector Machines* (2025). https://scikit-learn.org/stable/modules/svm.html
22. J. Wang, J. Yang, J. Zhang, X. Wang, W. Zhang, Big data driven cycle time parallel prediction for production planning in wafer manufacturing. Enterp. Inf. Syst. **12**(6), 714–732 (2018)
23. A.S. Ahmadian, Numerical modeling and simulation, in *Numerical Models for Submerged Breakwaters* (2016), pp. 109–126
24. T. Chen, The FLR-PCFI-RBF approach for accurate and precise WIP level forecasting. Int. J. Adv. Manuf. Technol. **63**, 1217–1226 (2012)
25. H. Yu, B.M. Wilamowski, Levenberg–Marquardt training, in *Intelligent Systems* (2018), pp. 12–1
26. T. Chen, An intelligent hybrid system for wafer lot output time prediction. Adv. Eng. Inform. **21**, 55–65 (2007)
27. T. Chen, A hybrid SOM-BPN approach to lot output time prediction in a wafer fab. Neural. Process. Lett. **24**(3), 271–288 (2006)
28. A.M. Kalteh, P. Hjorth, R. Berndtsson, Review of the self-organizing map (SOM) approach in water resources: analysis, modelling and application. Environ Model Softw. **23**(7), 835–845 (2008)
29. B. Waschneck, T. Altenmüller, T. Bauernhansl, A. Kyek, Production scheduling in complex job shops from an industry 4.0 perspective: a review and challenges in the semiconductor industry, in *SAMI@iKNOW* (2016), pp. 1–12
30. E.M. Kenny, M.T. Keane, Twin-systems to explain artificial neural networks using case-based reasoning: comparative tests of feature-weighting methods in ANN-CBR twins for XAI, in *Twenty-Eighth International Joint Conferences on Artificial Intelligence* (2019), pp. 2708–2715

31. T.C.T. Chen, Industry 4.0 for semiconductor manufacturing, in *Production Planning and Control in Semiconductor Manufacturing: Big Data Analytics and Industry 4.0 Applications* (2022), pp. 21–40
32. PEER Group Inc., SEMI PV2 (2022). https://www.peergroup.com/expertise/resources/semistandards/pv2/
33. S. Keil, Design of a cyber-physical production system for semiconductor manufacturing. Proc. Hamburg Int. Conf. Logistics **23**, 319–340 (2017)
34. J. Lee, B. Bagheri, H.A. Kao, Recent advances and trends of cyber-physical systems and big data analytics in industrial informatics, in *Proceeding of International Conference on Industrial Informatics* (2014), pp. 1–6
35. T.C.T. Chen, Y.C. Lin, Fuzzified deep neural network ensemble approach for estimating cycle time range. Appl. Soft Comput. **130**, 109697 (2022)
36. W. Fang, Y. Guo, W. Liao, K. Ramani, S. Huang, Big data driven jobs remaining time prediction in discrete manufacturing system: a deep learning-based approach. Int. J. Prod. Res. **58**(9), 2751–2766 (2020)
37. I. Tirkel, Cycle time prediction in wafer fabrication line by applying data mining methods, in *2011 IEEE/SEMI Advanced Semiconductor Manufacturing Conference* (2011), pp. 1–5
38. J. Choe, H. Shim, Attention-based dropout layer for weakly supervised object localization, in *Proceedings of the IEEE/CVF Conference on Computer Vision and Pattern Recognition* (2019), pp. 2219–2228
39. S. Ioffe, C. Szegedy, Batch normalization: Accelerating deep network training by reducing internal covariate shift, in *International Conference on Machine Learning* (2015), pp. 448–456
40. T. Chen, H.-C. Wu, M.-C. Chiu, A deep neural network with modified random forest incremental interpretation approach for diagnosing diabetes in smart healthcare. Appl. Soft Comput. **152**, 111183 (2024)
41. M.C. Chiu, T.C.T. Chen, Y.C. Wang, Flexible and objective diagnosis of type II diabetes by using a fuzzy deep learning ensemble approach. Complex Intell. Syst. **11**(6), 1–15 (2025)

Chapter 3
Cycle Time Modeling and Analysis

Abstract This chapter starts by defining the concept of cycle time modeling and analysis. Various cycle time modeling and analysis techniques and tools are then introduced. First, value stream mapping is applied to identify the operations of a job with the longest waiting times and should be improved first. Subsequently, after training the cycle time prediction model using machine learning or deep learning methods, a causal cycle time relationship analysis is performed to evaluate the impact of each job attribute on the cycle time forecast of a job, so as to find out the most important job attributes for the job. To this end, Shapely analysis and related tools are introduced with numerical examples and program codes. Subsequently, to explain and communicate the cycle time prediction mechanism and results, several explainable artificial intelligence (XAI) techniques and tools can be applied. First, for cycle time prediction problems involving big data, job classification is usually performed. Therefore, before predicting the cycle times, jobs need to be classified. To this end, traditional and XAI tools for explaining the job classification process and results are reviewed. Seven requirements that need to be met for excellent explanations are also listed. Subsequently, existing and XAI techniques and tools are applied to explain the cycle time prediction process and results, such as random forest-based incremental explanation. In addition, a systematic procedure is also established to confirm whether a trained cycle time prediction model conforms to domain knowledge, and on this basis, the improved random forest-based incremental explanation technique is proposed.

Keywords Cycle time modeling · Cycle time analysis · Random forest-based incremental interpretation · Domain knowledge · Explainable artificial intelligence

T.-C. T. Chen, *Advances in Cycle Time Management*,
https://doi.org/10.1007/978-3-032-06911-5_3

3.1 Cycle Time Modeling and Analysis

Cycle time modeling and analysis is to model the cycle time of a job, fit the relationship between job attributes and/or production conditions and cycle time, and analyze (or explain) the fitted relationship if it is generated by a deep learning (DL) model that is difficult to understand or communicate (see Fig. 3.1) [1].

The first cycle time modeling technique is fuzzy modeling. Due to the fluctuations in production conditions, the cycle time of a job is often uncertain and can be appropriately modeled using fuzzy logic. For example, a triangular fuzzy number (TFN) $\tilde{C}_j$ can be used to predict the cycle time of job j. Then, a set of linguistic terms is established to help evaluate whether the cycle time forecast is acceptable (see Fig. 3.2). In Fig. 3.2, the minimum and maximum cycle times are derived from the historical data.

If $\tilde{C}_j = (541, 549, 556)$, $\tilde{C}_j$ is mostly included in the linguistic term "short" (see Fig. 3.3), so the cycle time of this job is acceptable. In contrast,

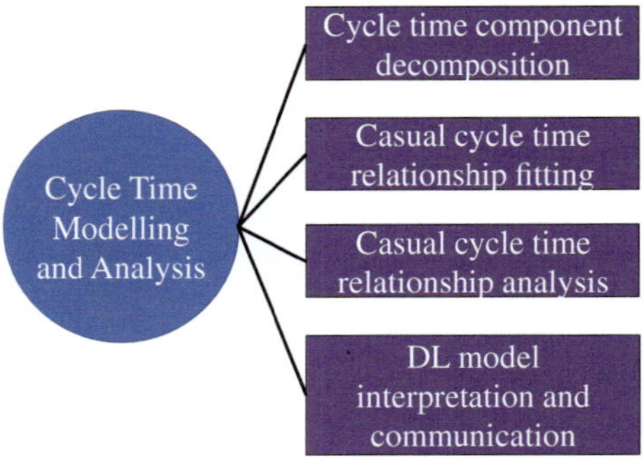

Fig. 3.1 Tasks in cycle time modeling and analysis

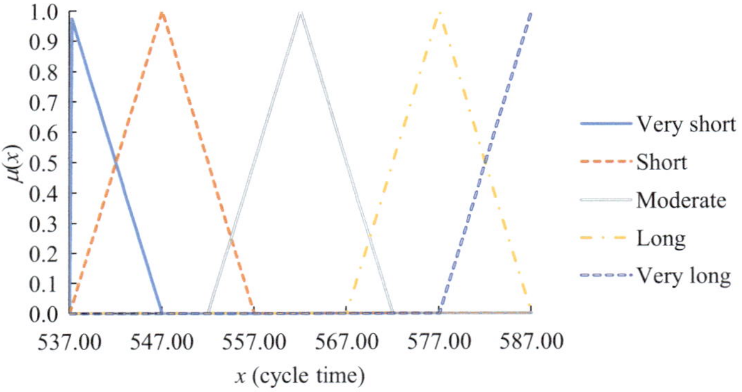

Fig. 3.2 Linguistic terms for evaluating whether a cycle time forecast is acceptable

if $\tilde{C}_j = (543, 556, 571)$, $\tilde{C}_j$ are mostly included in the two linguistic terms, "short" and "moderate" (see Fig. 3.4), so the cycle time of the job is only possibly acceptable.

Value stream mapping (VSM) [2] is another useful tool for cycle time modeling (see Fig. 3.5). From a VSM diagram, the waiting time percentage of each step can be easily derived:

$$w_{jk}\% = \frac{C_{jk} - p_{jk}}{p_{jk}} \cdot 100\% \tag{3.1}$$

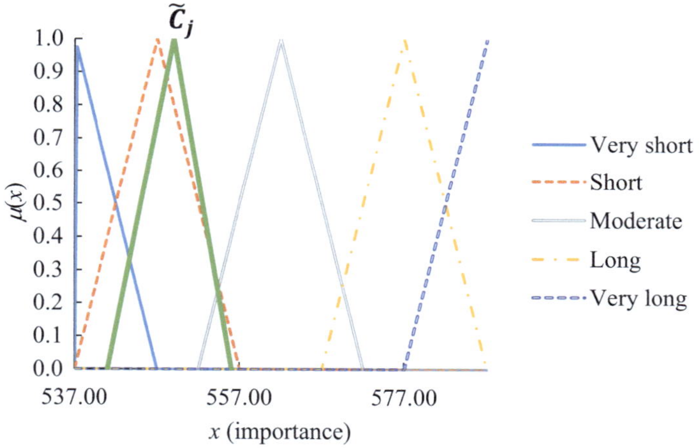

Fig. 3.3 Evaluation result for $\tilde{C}_j = (541, 549, 556)$

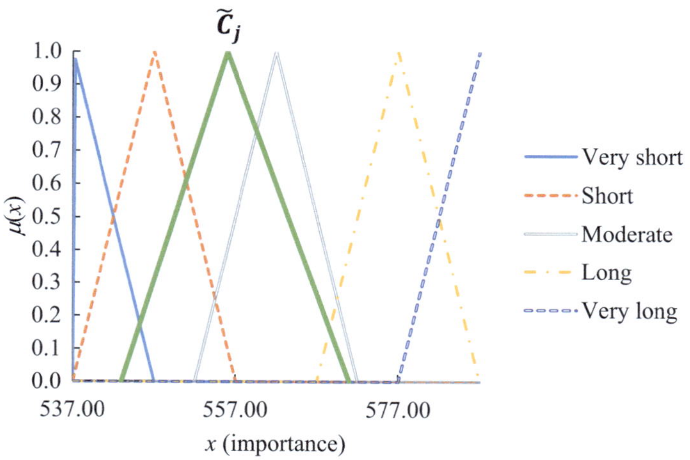

Fig. 3.4 Evaluation result for $\tilde{C}_j = (543, 556, 571)$

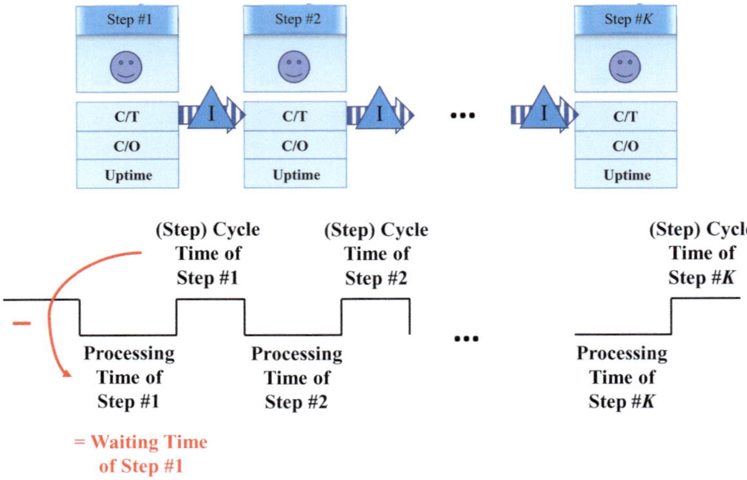

Fig. 3.5 VSM for cycle time modeling

Sorting all operations according to their waiting time percentages helps to identify the most urgent operations to be improved. Then, a waterfall plot can be drawn to sequence these steps according to their waiting time percentages to show their priorities to be improved, as illustrated in Fig. 3.6.

Example 3.1 There are four steps in manufacturing a product. A VSM diagram was drawn for cycle time modeling and improvement measure formulation (see Fig. 3.7). The waiting time percentages of these steps are calculated and ranked for prioritizing their improvement, as shown in Fig. 3.8.

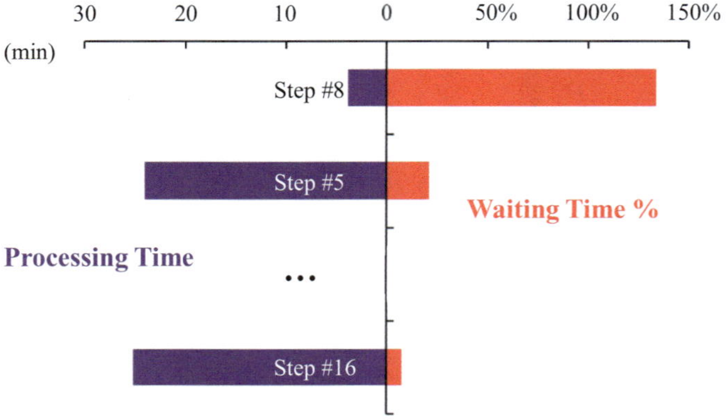

Fig. 3.6 Waterfall plot for sequencing the steps according to their waiting time percentages

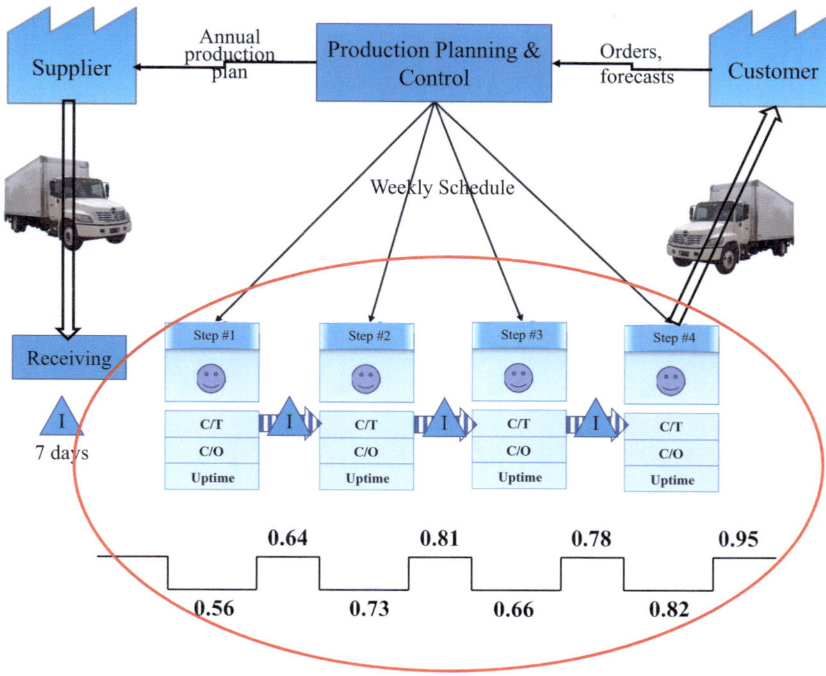

Fig. 3.7 VSM diagram for the product

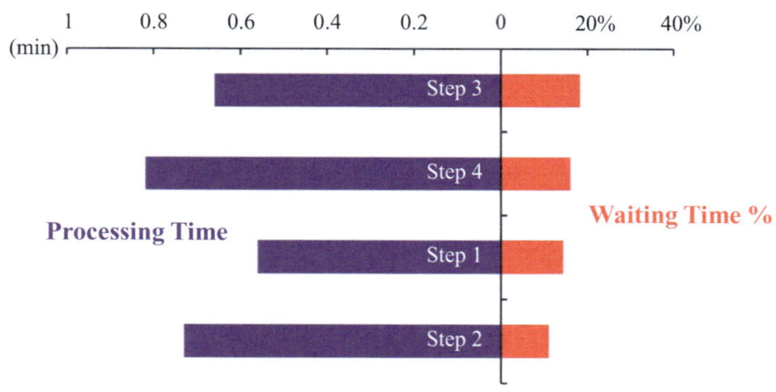

Fig. 3.8 Waterfall plot for prioritizing the improvement of these steps

This chapter focuses on causal cycle time relationship analysis and DL model interpretation and communication, as cycle time component decomposition and causal cycle time relationship fitting [3, 4] have been introduced in the previous two chapters.

3.2 Causal Cycle Time Relationship Analysis

After fitting the casual relationship between job attributes and/or production conditions and cycle time, the most important analysis is to find the factors that have the greatest impact on cycle time. In few cases, the fitted causal cycle time relationship is a simple model (i.e., a white box), such as linear regression (LR), and the conclusion is straightforward. However, in most cycle time prediction cases, DL methods are applied and the fitted causal cycle time relationship is very complicated and invisible (i.e., a black box). To overcome such difficulty, explainable artificial intelligence (XAI) comes to play, in which Shapley value (SHAP) analysis is probably the most popular technique [5].

3.2.1 *XAI Applications—SHAP Analysis*

The SHAP value is defined as the weighted average of the marginal contributions of all possible combinations $|\mathbf{F}|!$ as [5–7]:

$$\varphi_m(f) = \sum_{\{\mathbf{S}\subseteq\mathbf{F}\}\{m\}} \frac{|\mathbf{S}|!(|\mathbf{F}| - |\mathbf{S}| - 1)!}{|\mathbf{F}|!} \circ \left[f\left(x_{\mathbf{S}\cup\{m\}}\right) - f\left(x_{\mathbf{S}}\right)\right] \tag{3.2}$$

in which

- $\varphi_m(f)$ is the weighted average Shapley value that factor (job attribute or production condition) m provides in the context of all combinations that exclude factor m.
- $\mathbf{F}$ is the set of all factors.
- $\mathbf{S}$ is a subset (i.e., coalition) of $\mathbf{F}$.
- $f\left(x_{\mathbf{S}\cup\{m\}}\right)$ is the cycle time forecast generated by the DL model considering factor m, while $f\left(x_{\mathbf{S}}\right)$ is the cycle time forecast without considering factor m.

When $|\mathbf{S}| = |\mathbf{F}| - 1$, i.e., $\mathbf{S} = \mathbf{F} \setminus \{m\}$, $|\mathbf{S}|!(|\mathbf{F}| - |\mathbf{S}| - 1)! = |\mathbf{F}|!$, then

$$\varphi_m(f) = \sum_{\{\mathbf{S}\subseteq\mathbf{F}\}\{m\}} \frac{\left[f\left(x_{\mathbf{S}\cup\{m\}}\right) - f\left(x_{\mathbf{S}}\right)\right]}{|\mathbf{F}|!} \tag{3.3}$$

In this way, SHAP evaluates the importance of a factor by fixing its value while randomizing the values of other inputs [8], and then averaging the resulting differences in $f()$. However, sometimes certain factors cannot be meaningfully randomized. SHAP analysis results are often summarized with a waterfall (or tornado) plot (see Fig. 3.9). As can be seen from the figure, the cycle time forecast of the job is 1298.39 (hrs). The absolute value of the SHAP value of x_{j2} is the highest, so x_{j2} is the most important factor to the cycle time forecast of the job, followed by x_{j4}. However, the SHAP value of x_{j2} is positive, which means that setting x_{j2} to the current value

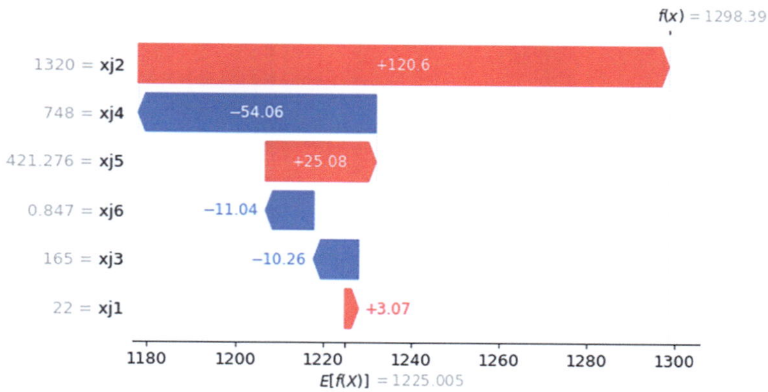

Fig. 3.9 Waterfall plot for summarizing SHAP analysis results [5]

increases the cycle time forecast by 120.6 (hrs), which is not a conducive property to cycle time management.

Another visualization tool that serves a similar purpose to the waterfall plot is a force plot that is more compact (see Fig. 3.10) [9]. One slight difference is that a force plot shows the value, rather than the contribution, of each factor for the job. Nonetheless, as in a waterfall plot, the width of each bar corresponding to a factor is still proportional to the contribution of the factor. For example, according to this figure, setting x_{j2} to 1320.0 increases the cycle time forecast of the job by 120.6 (hrs). In contrast, setting x_{j4} to 748.0 decreases the cycle time forecast by 54.06 (hrs).

Subsequently, a scatter plot can be used to evaluate and summarize the importance of a factor to all jobs (see Fig. 3.11). Take x_{j1} as an example. The factor having the highest correlation with x_{j1} is x_{j3}. The line on the right side of Fig. 3.11 indicates that the color in the figure represents the value of x_{j3} from small (blue) to large (red). From the figure, it can be seen that

- As the value of x_{j1} increases, the SHAP value of x_{j1} also increases.
- In addition, the larger the value of x_{j3}, the higher the SHAP value of x_{j1}.

Finally, a bee swarm plot can be drawn to summarize the importance of all factors for all jobs, as shown in Fig. 3.12. As can be seen from the figure,

- x_{j2} has the highest importance to the cycle time forecasts of all jobs.

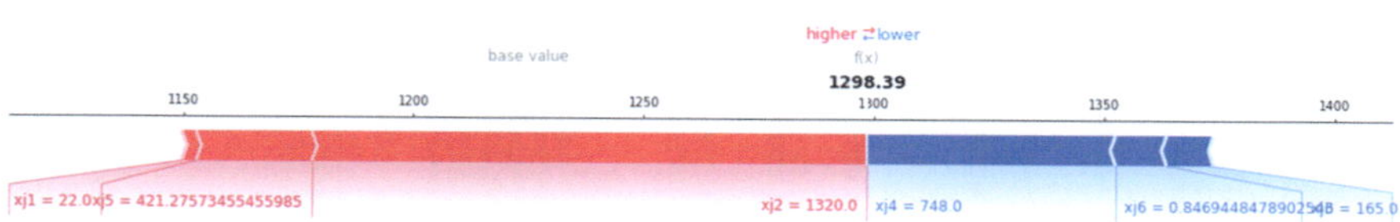

Fig. 3.10 Force plot

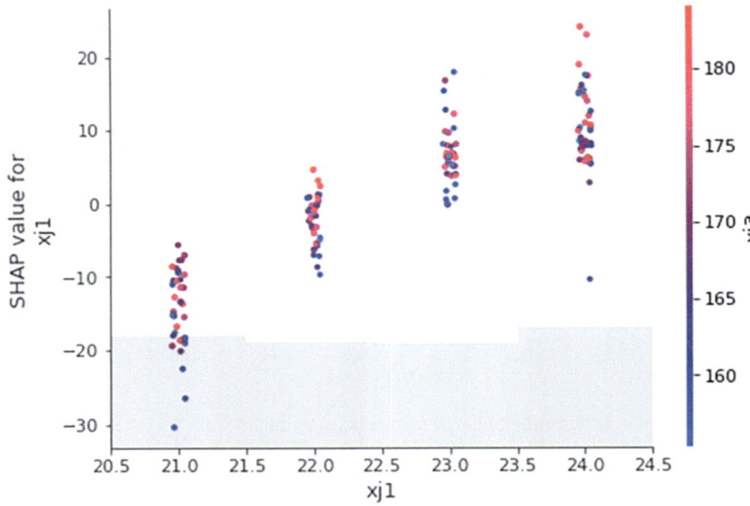

Fig. 3.11 Scatter plot for x_{j1}

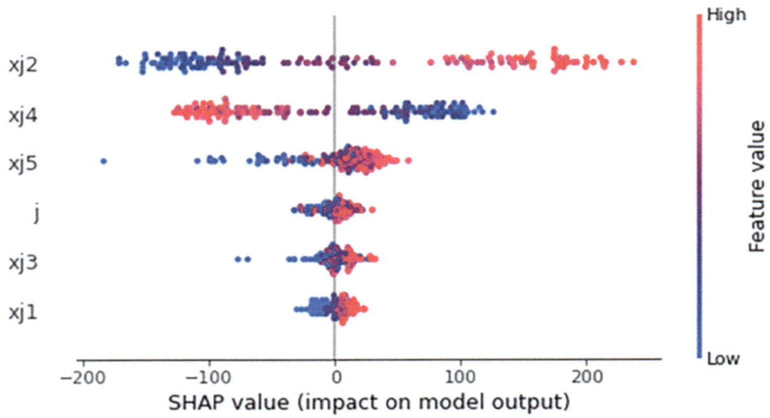

Fig. 3.12 Bee swam plot

- However, the range of the SHAP values of x_{j2} is the widest, there is a lot of uncertainty.
- Judging from the colors of points, basically, the SHAP value of x_{j2} will be higher when the value of x_{j2} increases.

Example 3.2 Cycle time-related data have been collected for 200 jobs (see Table 3.1), in which the cycle time of a job is predicted based on its six factors (see Table 3.2). The data of the first 150 jobs are used as the training data, while the data of the remaining 50 jobs are reserved for testing. eXtreme Gradient Boosting (XGBoost) is applied to predict the cycle times of jobs. Then, SHAP analysis is performed to

Table 3.1 Cycle time-related data of 200 jobs

j	x_{j1}	x_{j2}	x_{j3}	x_{j4}	x_{j5}	x_{j6} (%)	C_j (hrs)
1	22	1320	165	748	421	85	1298
2	23	1261	169	734	198	81	1098
3	22	1187	159	635	287	93	1122
…							
200	24	1309	175	734	345	81	1276

Table 3.2 Factors

x_{j1}	x_{j2}	x_{j3}	x_{j4}	x_{j5}	x_{j6}
Job size	Factory work-in-process (WIP)	Queue length before bottlenecks	Factory queue	Average waiting time of recently completed jobs	Factory utilization

assess the importance of each factor to the cycle time forecast of a job. The required Python code is provided in Fig. 3.13. The SHAP analysis results for the first job in test data (Job #151) are summarized with a waterfall plot in Fig. 3.14. According to this figure, in this example, x_{j2}, factory work-in-process (WIP), is the most important factor to the cycle time forecast of this job. The force plot for this job is presented in Fig. 3.15.

Example 3.3 Taking x_{j2}, factory WIP, as an example, a scatter plot is also drawn for Example 3.2, as shown in Fig. 3.16. The factor having the highest correlation with x_{j2} is x_{j5}, the average waiting time of recently completed jobs. From the figure, it can be seen that as the value of x_{j2} increases, the SHAP value of x_{j2} also increases. In addition, a lower SHAP value of x_{j2} is associated with a lower value of x_{j5}, and vice versa.

Example 3.4 A bee swarm plot is also drawn to summarize the importance of all factors for all jobs in Example 3.2, as shown in Fig. 3.17. As can be seen from the figure, x_{j2} has the highest importance to the cycle time forecasts of all jobs. However, the range of the SHAP values of x_{j2} is the widest, there is a lot of uncertainty. Another obvious trend is that as the value of x_{j2} increases, the SHAP value of x_{j2} also increases.

3.3 Interpretation and Communication of Cycle Time Prediction Mechanism

In recent years, many DL methods have been proposed to enhance the accuracy of predicting the cycle times of jobs. However, the prediction mechanism of a DL application is difficult to understand and communicate to users [10], which limits its

```
import xgboost
import shap
import pandas as pd
import numpy as np

# Import the collected data
cycletimedata = pd.read_excel('F:\\ACTM\\example2.xlsx')
# Split the data
train_x=cycletimedata.iloc[0:150,0:6]
train_y=cycletimedata.iloc[0:150,7]
test_x=cycletimedata.iloc[151:200,0:6]
test_y=cycletimedata.iloc[151:200,7]
train_y=train_y.to_numpy()

# Train the forecasting model
model1=xgboost.XGBRegressor().fit(train_x, train_y)

# SHAP analysis
explainer1=shap.Explainer(model1)
shap_values1=explainer1(test_x)

# Show the analysis results for the first job in test data
shap.plots.waterfall(shap_values1[0])
shap.force_plot(shap_values1[0], matplotlib=True)

# Analyze the scatter of SHAP values of xj1
shap.plots.scatter(shap_values1[:, "xj1"], color=shap_values1)

# Analyze the scatter of SHAP values of all factors
shap.plots.beeswarm(shap_values1)
```

Fig. 3.13 Python code for SHAP analysis

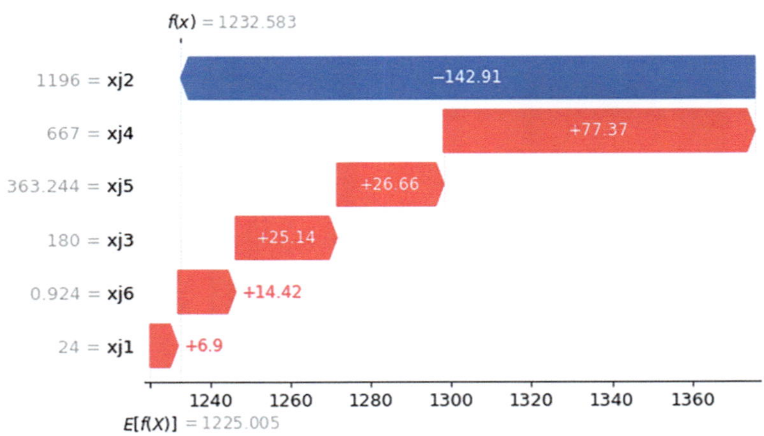

Fig. 3.14 SHAP analysis results for the first job in test data

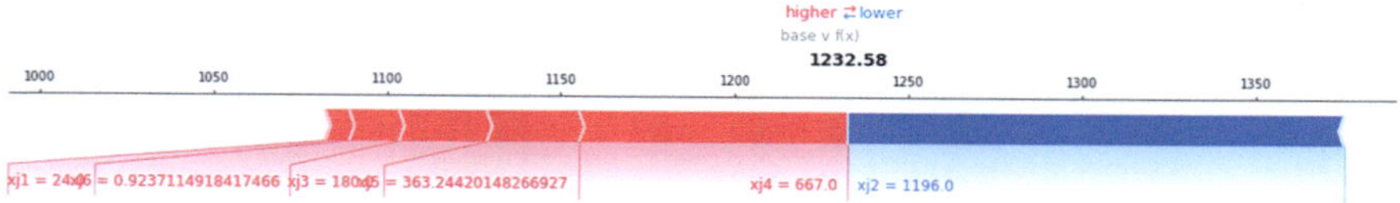

Fig. 3.15 Force plot for the job

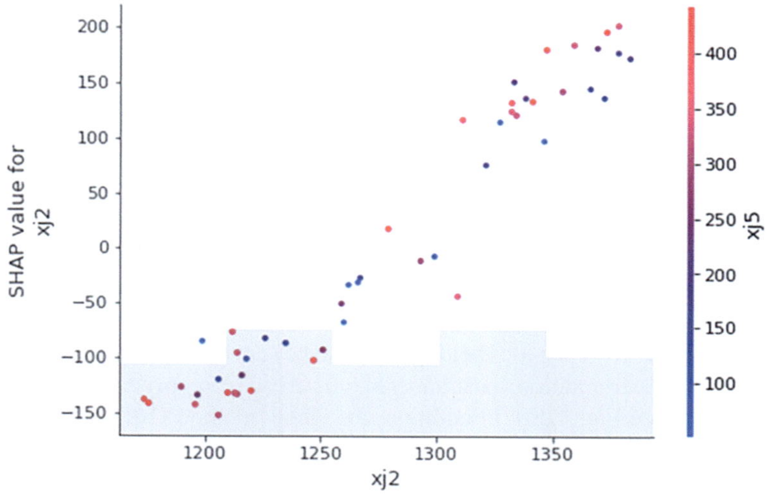

Fig. 3.16 Scatter plot for Example 3.1

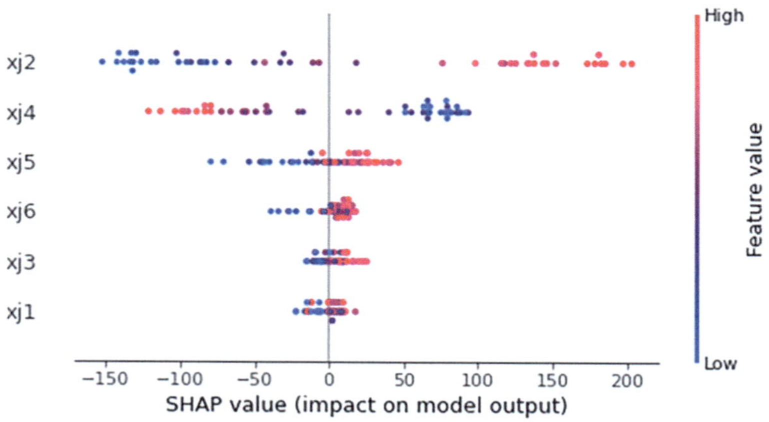

Fig. 3.17 Bee swam plot for Example 3.2

acceptability (or practicality). As described by Chen and Wang [11], most state-of-the-art cycle time prediction methods classify jobs before predicting their cycle times. Therefore, the same procedure is followed when explaining and communicating the mechanism and results of a cycle time prediction method as follows.

3.3.1 Explaining Job Classification Process and Results

Traditional techniques such as text descriptions, flowcharts, and pseudocodes have been widely used to explain job classification process [12–14]. In addition, animation-based XAI techniques such as ConvNetJS [15] have also been used to illustrate job classification process and results using artificial neural networks (ANNs) or deep neural networks (DNNs).

When there are only two attributes, the classification results can be visualized using a scatter plot [11], as shown in Fig. 3.18, where jobs are divided into three categories: x_{jp} is attribute p of job j; $p = 1 \sim 2$; $j = 1 \sim n$. The classification results (i.e., the categories to which jobs belong) can also be summarized with a line chart, as shown in Fig. 3.19. In addition, the number of jobs in each category can be represented by a bar chart (see Fig. 3.20). In addition, the elbow method [16] describes a way to determine the optimal number of job categories (see Fig. 3.21), from which the reduction in inertia becomes negligible.

When the number of attributes is large (i.e., high-dimensional), self-organization maps (SOMs) provide a graphical tool to display the distance between adjacent job categories, as shown in Fig. 3.22. Job classifications are represented by gray-blue blocks and connected to their adjacent categories by red lines. The color of the connecting blocks ranges from black to yellow, indicating the distance between two job categories. The darker the color of the connecting block, the greater the distance.

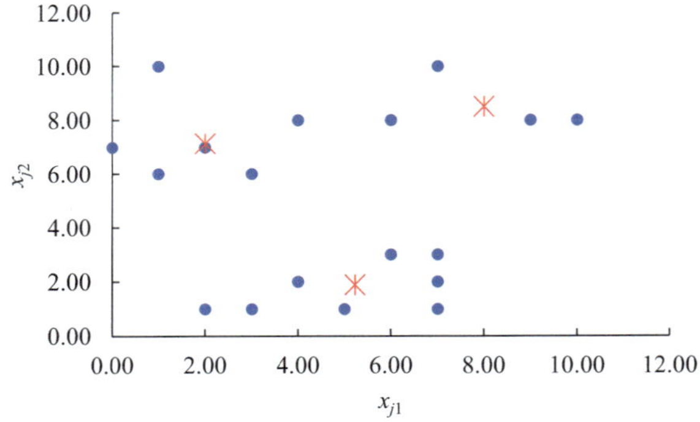

Fig. 3.18 Visualizing the classification results using a scatter plot [11]

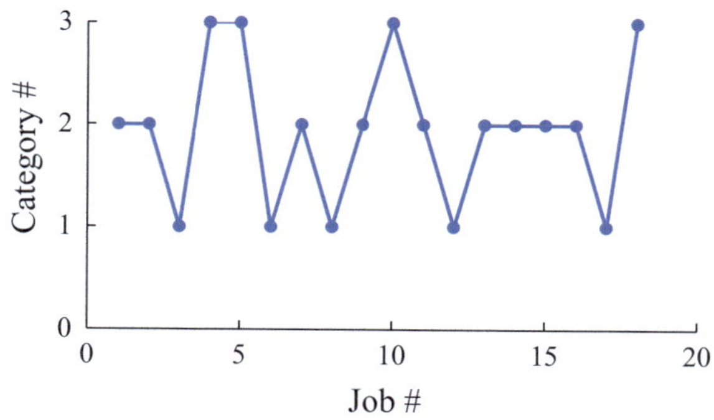

Fig. 3.19 Summarizing the classification results with a line chart

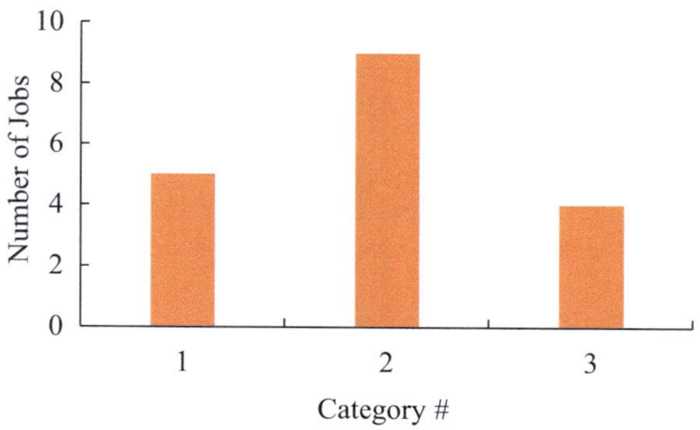

Fig. 3.20 Bar chart showing the number of jobs in each category

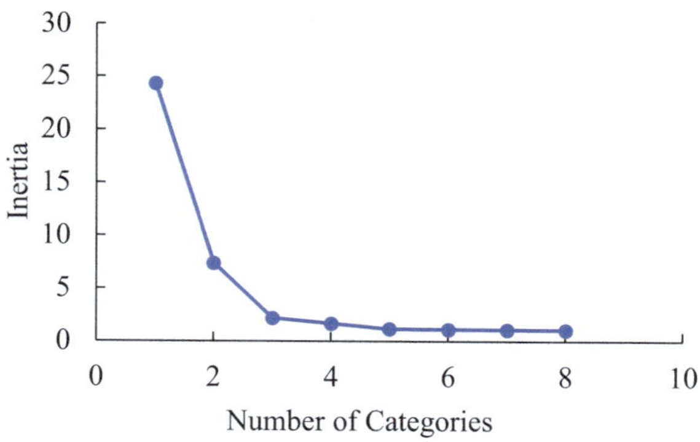

Fig. 3.21 Line chart for determining the optimal number of job categories using the elbow method

3.3.2 *XAI Applications—Saliency Maps*

Saliency maps are images that highlight the areas where people's eyes focus first or are areas that are most relevant to an ML or DL application. The initial scope of a saliency map is to mark the input-to-output relationship, that is, to distinguish the importance of input pixels. In Fig. 3.21, the number of categories is regarded as

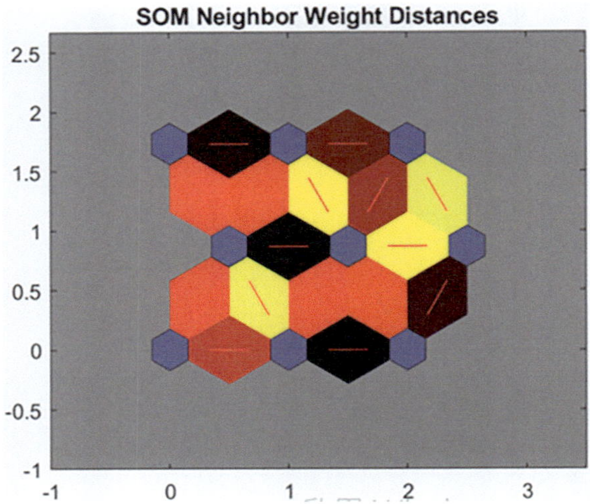

Fig. 3.22 SOM for visualizing the distance between adjacent job categories [11]

the input, and the global optimality of the inertia is regarded as the output. In this way, the saliency map can be used to detect the convergence of the inertia reduction process.

The distances of each pixel to the rest of the pixels in the frame are calculated and summed first to build a saliency map for a frame according to the following expression:

$$\mathrm{SALS}(I_k) = \sum_{i=1}^{n} |I_k - I_i| \tag{3.4}$$

where $I_i \in [0, 255]$ is the value of pixel k, i and n is the number of pixels in the frame. To highlight pixels from others:

$$\mathrm{SALS}(I_k) = \sum_{l} F_l |I_k - I_l| \tag{3.5}$$

where F_l is the frequency of I_l. $l \in [0, 255]$.

Chen et al. [17] applied the concept of saliency maps to detect the convergence of the inertia reduction process, in which data points are associated with different colors according to their performances in reducing inertia:

$$\left[R_{p_t}, G_{p_t}, B_{p_t}\right] = \left[255 \cdot \left(1 - \frac{p_t - \min\limits_s p_s}{\max\limits_s p_s - \min\limits_s p_s}\right), 255 \cdot \frac{p_t - \min\limits_s p_s}{\max\limits_s p_s - \min\limits_s p_s}, 0\right] \tag{3.6}$$

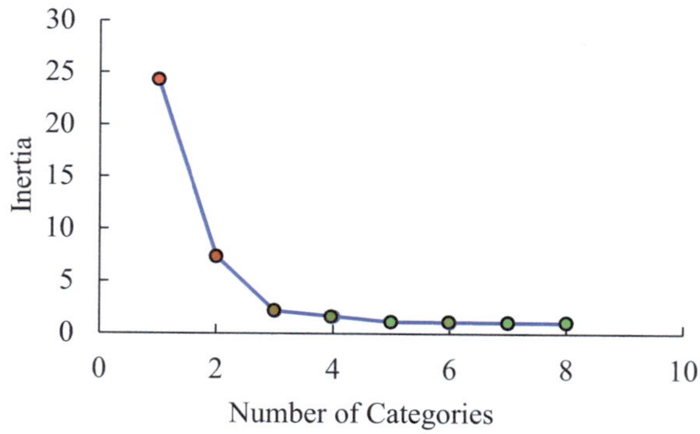

Fig. 3.23 Saliency maps application to Fig. 3.21

where p_t is the t-th data point, and R_{p_t}, G_{p_t}, and B_{p_t} are the red, green, and blue pixels of p_t, respectively. Compared with the original formula for evaluating saliency, Eq. (3.4) estimates the normalized value rather than the absolute deviation because the p_t value may be very large. An example is given in Fig. 3.23, in which green points are associated with better performances than red points. Thus, when a data point turns green, it indicates that the performance is above the average. Whether the evolution process needs to be terminated depends on how green the current data point is for the analyst.

3.3.3 Comparison of Various Explanation Techniques

Chen and Wang [11] made a table for comparing existing techniques for explaining job classification process and results (see Table 3.3). A good explanation technique needs to meet seven requirements, while most existing explanation techniques can only meet five requirements at most. Obviously, there is no perfect explanation technique that can explain job classification process and results. Therefore, existing explanation techniques should be improved or new (or customized) explanation techniques can be proposed.

3.3.4 Interpreting Cycle Time Prediction Process and Results

Since the relationship between job attributes (or production conditions) and cycle time is usually nonlinear, ANNs and DNNs have been widely used for cycle time prediction. Text descriptions and network configuration (or system architecture) diagrams are frequently applied to explain the operations in ANNs (or DNNs) [18]. In addition, animation-based techniques, such as ConvNetJS (for convolutional neural networks, CNNs) [15], TensorFlow (for ANNs, DNNs with multiple hidden layers)

Table 3.3 Comparison of existing techniques for explaining job classification process and results

	I	II	III	IV	V	VI	VII	VIII	IX
Easy to understand	V	V		V	V	V			V
Easy to communicate		V		V	V	V		V	V
Show the classification process	V	V	V						V
Show the relative position of a job among categories				V					V
Provide statistics on categories				V	V	V	V	V	
No background knowledge required	V	V		V	V	V			
Can handle high-dimensional data								V	

I: Textual description; II: Flowchart; III: Pseudocode; IV: Scatter plot; V: Line chart (Fig. 3.21); VI: Bar chart; VII: Line chart (Fig. 3.23); VIII: SOM; IX: ConvNetJS

[19], and MATLAB [20], have been used to illustrate the training process of ANNs or DNNs.

To better explain the results of cycle time prediction using ANNs, Green et al. [21] evaluated the influence of each job attribute on the cycle time forecast by taking the partial derivative of the cycle time forecast with respect to the job attribute. The result included exponential functions and inverse functions, but applied to all jobs, and therefore served as a global explanation [22]. This result could also be used to derive Taylor series to approximate ANNs. To generate local explanations, Green et al. calculated the odd ratio of each job attribute. The job attribute with the highest odd ratio was considered to be the most influential attribute for the job.

Kenny and Keane [23] built a case-based reasoning (CBR) system to approximate the operations in an ANN for job cycle time prediction. After training the ANN, synthetic jobs [24] were randomly generated and fed into the ANN. Then, k-means [25] was applied to extract cases from these synthetic jobs. To predict the cycle time of job j, the distance between job j and case k was measured. Then, the cycle time of this job was predicted with the weighted average of the cycle times of all cases, where the weight was the inverse of the corresponding distance. Only the inputs and output of the ANN are considered in the rules of CBR, i.e., a post hoc XAI method. Therefore, CBR can be applied to approximate operations in more complex ANNs and DNNs, such as ANNs with multiple hidden layers, recurrent neural networks, and fuzzy neural networks. For example, Kenny and Keane [23] built another CBR system to approximate a CNN. However, the rule content in such a CBR system was usually too simple to provide accurate approximations.

Classification and regression trees (CARTs) have also been constructed to approximate ANNs or DNNs for cycle time prediction using decision rules [26]. All job attributes do not have to appear in each rule. In addition, the order in which job attributes appear in different rules may not be the same. The steps for constructing a CART can be found in [26].

The above methods suffer from low approximation accuracy. A random forest (RF) is a collection of classification and decision trees that consider different parts of the data and try various combinations of job attributes [27]. In this way, RF is usually

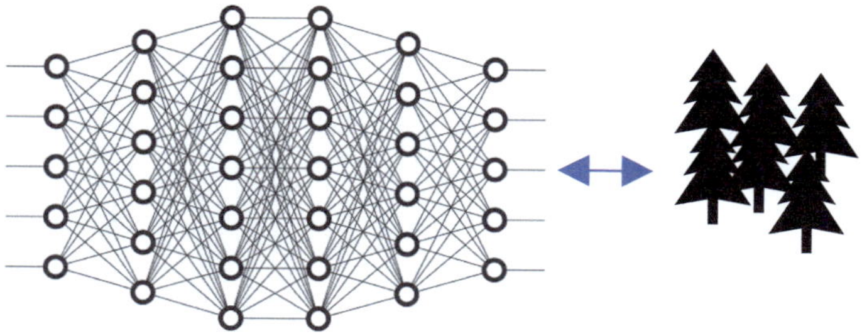

Fig. 3.24 RF for approximating DNN operations

more accurate than a single CART model and is expected to better approximate an ANN or DNN. With RF, the cycle time forecast made by an ANN (or DNN) for job j is predicted by a rule in each tree. Cycle time forecasts by all trees are then averaged. However, using multiple rules to approximate the cycle time forecast by the ANN (or DNN) for a job may confuse the user.

Chen and Lin [28] proposed a hybrid XAI–fuzzy collaborative intelligence (FCI) approach for estimating the job cycle time range. In the XAI–FCI approach, experts collaborated to predict the job cycle time using DL models. Then, each DL model was explained by a RF (see Fig. 3.24), and the RFs of all DL models collaborated to fuzzify the cycle time forecast to estimate the job cycle time range. Similar treatments have been tried in recent studies [29, 30]. An RF itself can be applied to cycle time prediction [29]. Therefore, if the accuracy of cycle time prediction using an RF is better than that using a DL model, the RF should be applied directly. Otherwise, a DL model should be applied first, and then RF can be used to explain the DL application to enhance its interpretability and communicability.

Table 3.4 compares existing techniques for interpreting ANNs (or DNNs) for cycle time prediction. An excellent interpretation technique needs to meet seven requirements. Among the existing interpretation techniques, RF meets most of the requirements.

3.3.5 *Big Data Analytics and XAI Application*

In the big data analytics application for explaining job classification process and results, first, k-means is applied to classify jobs [11]. The procedure for k-means comprises the following steps:

Step 1. Determine the number of categories K.
Step 2. Randomly assign jobs to categories.
Step 3. Calculate the center of each job category:

Table 3.4 Comparison of existing interpretation techniques for explaining ANNs or DNNs for cycle time prediction

	I	II	III	IV	V	VI	VII	VIII	IX
Easy to understand	V	V		V	V		V	V	V
Easy to communicate		V	V	V	V		V	V	V
Show the training process	V		V	V	V				
Show a direct relationship between job attributes and cycle time forecast						V	V	V	V
Rules are simple							V	V	
No background knowledge required	V						V	V	V
High approximation accuracy			V	V	V				V

I: Textual description; II: Network configuration diagram; III: ConvNetJS; IV: TensorFlow; V: MATLAB; VI: Partial derivation/Odd ratio; VII: CBR; VIII: CART; IX: RF

$$c_{kp} = \frac{\sum_{j \in \mathbf{U_k}} x_{jp}}{\sum_{j \in \mathbf{U_k}} 1} \tag{3.7}$$

where $\mathbf{U_k}$ represents job category k; $k = 1 \sim K$. From the perspective of XAI, $\mathbf{x_j} = \{x_{jp}\}$ can be visualized using a radar chart to address the big data issue (see Fig. 3.25). $\mathbf{c_k} = \{c_{kp}\}$ is the center of job category k. $\mathbf{c_k}$ can also be visualized using a radar chart.

Step 4. Measure the distance between a job and the center of each category as

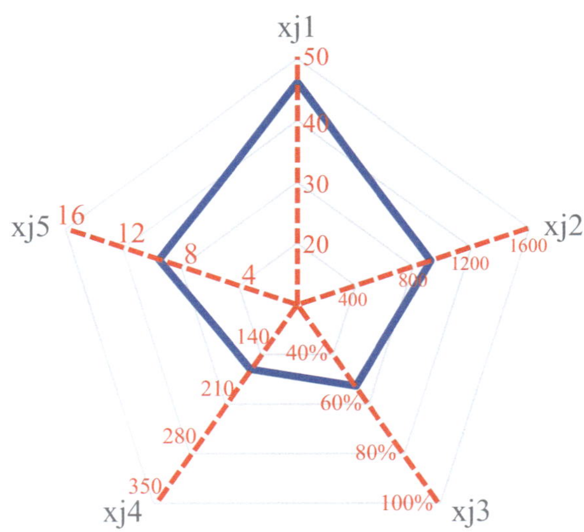

Fig. 3.25 Radar chart for visualizing the high-dimensional data of a job

$$d_{jk} = \sqrt{\sum_{p=1}^{P} (x_{jp} - c_{kp})^2} \tag{3.8}$$

where d_{jk} is the distance between job j and the center of category k.

Step 5. Reassign each job to the nearest category.

Step 6. Calculate the sum of within-category squared distances (SSD) as

$$Z_1 = \sum_{k=1}^{K} \sum_{j \in \mathbf{U_k}} d_{jk}^2 \tag{3.9}$$

Step 7. If Z_1 converges, stop; otherwise, return to Step 4.

When the cycle time of a new job is to be predicted, its data is compared with the centers of all categories. Then, the job is assigned to the nearest category. The comparison result can also be better visualized using a scatter radar diagram to address the big data issue, as illustrated in Fig. 3.26. In the scatter radar diagram,

- The data of a new job and the centers of all categories (or the closest categories) are presented in the form of radar charts.
- The job is placed at the center.
- The centers of categories closer to the job are placed around the job.
- A directional arrow links the job to the center of each category.
- A thicker arrow indicates that the job is closer to the center of the category.

Example 3.5 The cycle times of jobs in a dynamic random access memory (DRAM) manufacturing factory are to be predicted. Six production conditions and job attributes are considered in predicting the cycle time of a job: job size (x_{j1}), factory work-in-process (WIP) (x_{j2}), the total queue length before bottlenecks (x_{j3}), factory total queue length (x_{j4}), the average waiting time of recently completed jobs (x_{j5}), and factory utilization (x_{j6}). The data of 120 jobs have been collected. The first 80 jobs are used to construct the prediction model, while the remaining jobs are left for testing/evaluation. Take job #1 as an example. Its (normalized) data is visualized using a radar scatter chart in Fig. 3.27. The common expression technique [31] is also applied to replace variables and symbols with common expressions. Obviously, the size and factory queue length of job #1 are relatively large among all jobs.

Example 3.6 In Example 3.5, jobs are classified into five categories using k-means. The classification results are summarized with a line chart in Fig. 3.28. In addition, the number of examples in each category is shown by a bar chart in Fig. 3.29.

Example 3.7 In Example 3.5, the relative position of job #1 among all categories is illustrated using a scatter radar diagram in Fig. 3.30. Job #1 is assigned to category #2, but is also close to categories #4 and #5.

As a new XAI technique for explaining the classification result, the scatter radar diagram meets the following requirements:

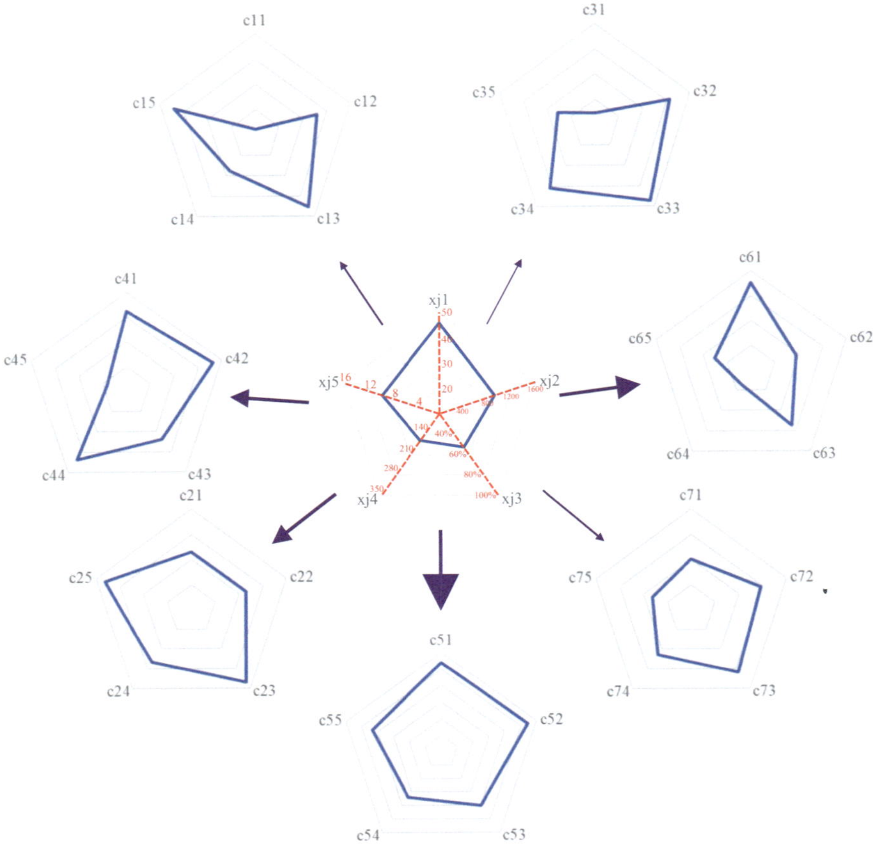

Fig. 3.26 Visualizing the classification result of a new job using a radar scatter diagram

Fig. 3.27 Visualizing the data of job #1

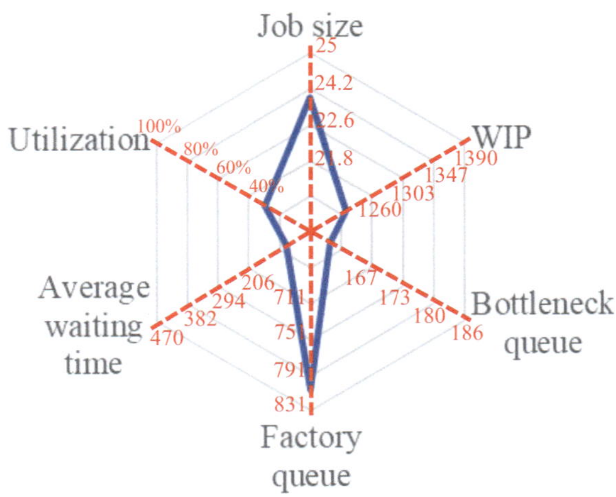

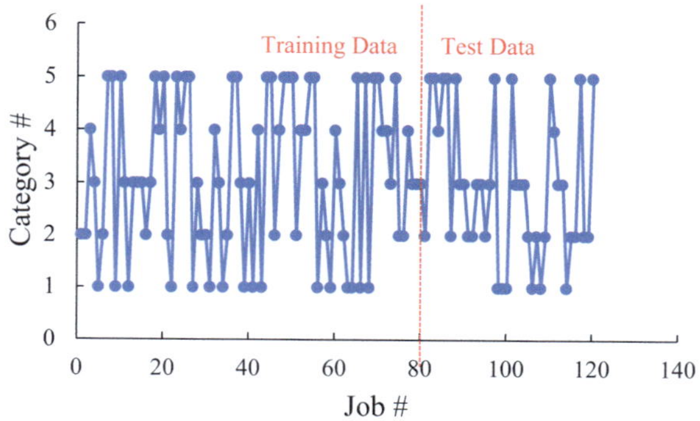

Fig. 3.28 Classification results using k-means

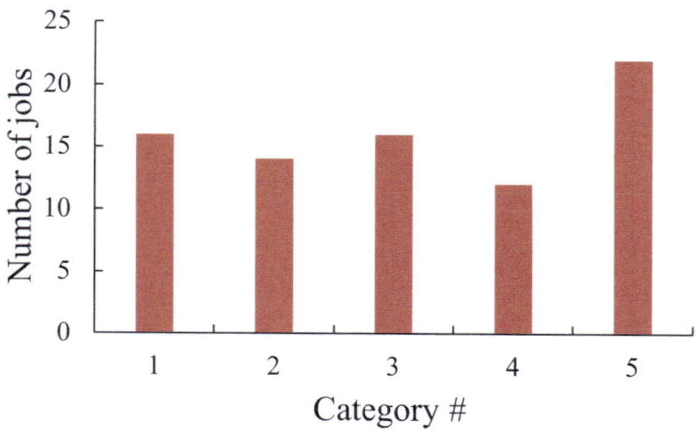

Fig. 3.29 Number of jobs in each category (training data)

- It is easy to understand.
- It is easy to communicate.
- It shows the relative position of a job among categories by differentiating the thicknesses among arrows.
- It provides statistics on job categories, i.e., the centroid of each job category.
- It does not need background knowledge.
- It can handle high-dimensional job data.

Then, the ANN (or DNN) of the category to which the job belongs is applied to predict the cycle time of the job.

3.3.6 XAI Application

However, using the centroid of each category as the representative value to predict the cycle times of all jobs in the category may be far from accurate. Therefore, subsequently, an ANN (or DNN) is built for each category to predict the cycle times of jobs in that category. However, it is very difficult to express the relationship between

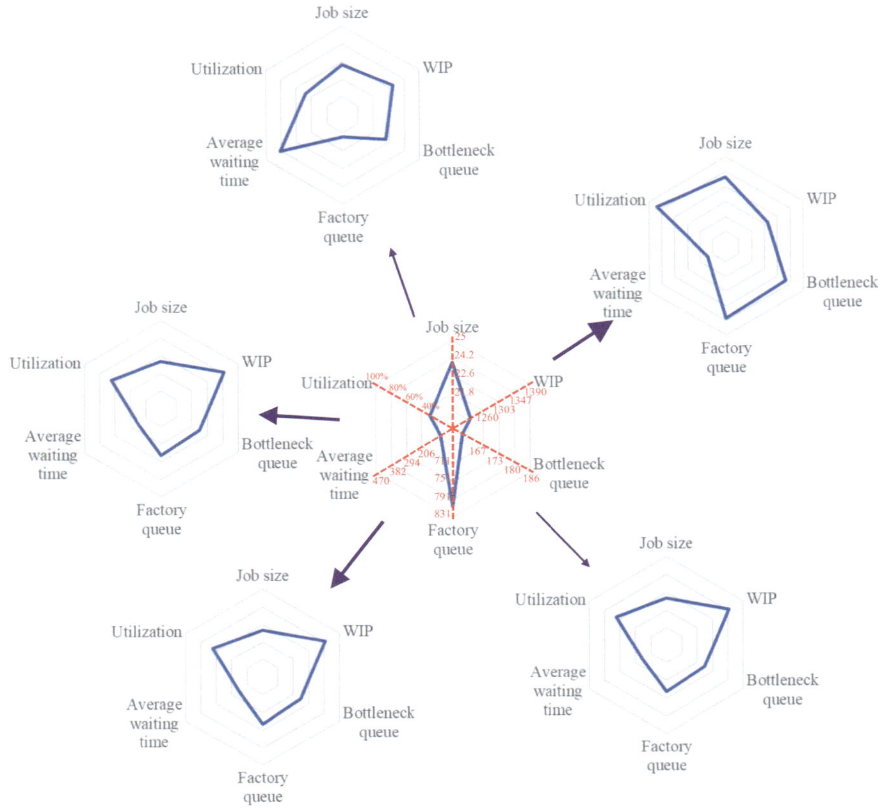

Fig. 3.30 Scatter radar diagram for illustrating the relative position of job #1 among all categories

the inputs and output of the ANN (or DNN) with a simple model. To overcome this difficulty, RF is applied because it can achieve higher approximation accuracy in terms of the sum of squared error (SSE) than other traditional techniques such as multiple linear regression, CBR, and CART:

$$\mathrm{SSE} = \sum_{j=1}^{n} \left(C_j - \hat{C}_j \right)^2 \tag{3.10}$$

where $\hat{C}_j$ is the approximated cycle time forecast of job j.

Example 3.8 In Example 3.5, an ANN is constructed for each job category to predict the cycle times of jobs belonging to the category. The ANN has three layers: the input layer, a single hidden layer with 12 nodes, and the output layer. The ANN is trained using the LM algorithm. It is noteworthy that the optimization of the ANN is not the focus. Taking job category #1 as an example. The cycle time forecasting results

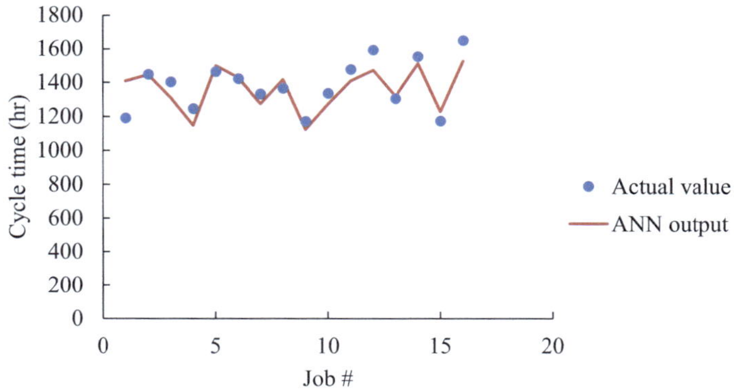

Fig. 3.31 Forecasting results (job category #1 in the training data)

are summarized in Fig. 3.31. The forecasting performance (by applying the trained ANN to test data), measured in terms of MAE, MAPE, and RMSE, is

$$\text{MAE} = 170\text{ hr}$$
$$\text{MAPE} = 11.9\%$$
$$\text{RMSE} = 208\text{ hr}$$

Example 3.9 Following Example 3.8, an RF [32] is constructed to approximate the cycle time prediction mechanism and results using the ANN. The RF is composed of three decision trees (see Fig. 3.32). It is worth noting that after classification, one job category may have only a few jobs, so the ANN as well as the approximating RF becomes less complex. The approximation results are summarized in Fig. 3.33.

However, an RF consists of multiple trees. From each tree, a decision rule can be applied to predict the cycle time of a job. As a result, multiple rules are applicable at the same time, which may cause confusion to the user. To address this issue, Chen and Wang proposed random forest-based incremental interpretation (RFII) that re-organized these decision rules as follows:

Step 1. Average the prediction results by decision rules as $\hat{C}_j$:

$$\hat{C}_j = \frac{\sum_{t=1}^{T} \hat{C}(t)}{T} \tag{3.11}$$

Step 2. Set $q = 1$.
Step 3. Choose the decision rule with the closest prediction result to $\hat{C}_j$.
Step 4. Set the q-th RFII rule to the decision rule.
Step 5. Set the current approximated output $\hat{C}_j$(current) to the prediction result of the decision rule.
Step 6. Remove the decision rule.

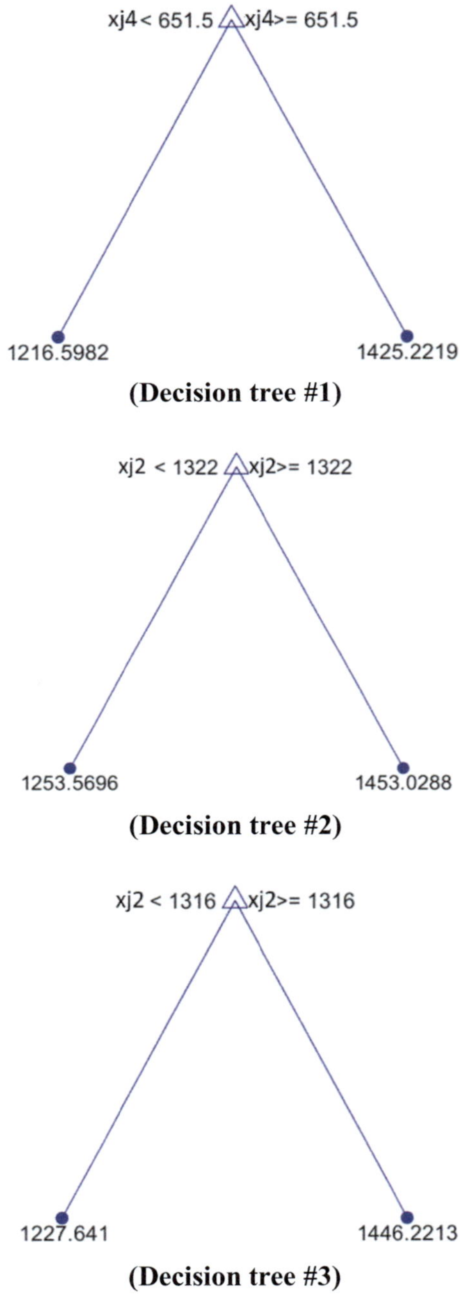

Fig. 3.32 Decision trees of the RF

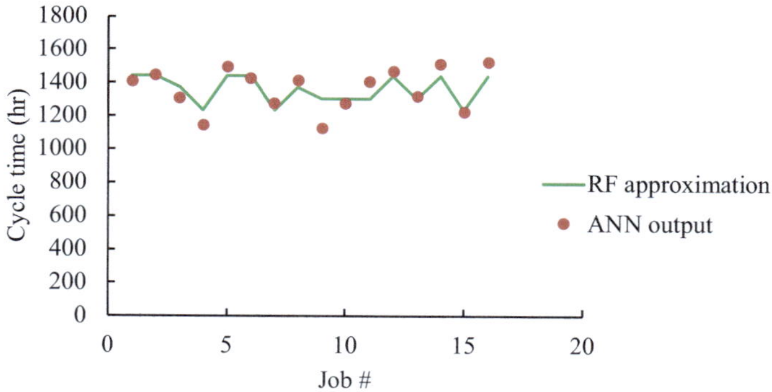

Fig. 3.33 RF approximation results

Step 7. Increase q by 1.
Step 8. If $q > T$, go to Step 14; otherwise, go to Step 9.
Step 9. Find the decision rule with the closest prediction result to $\hat{C}_j$.
Step 10. Subtract $\hat{C}_j$(current) from the prediction result of the decision rule, and divide the result by T: The result is indicated by $\Delta\hat{C}_j$(current).
Step 11. Set the premise of the q-th re-organized to that of the decision rule.
Step 12. Set the consequence of the q-th RFII rule to "add $\hat{C}_j$ by $\Delta\hat{C}_j$(current)."
Step 13. Return to Step 6.
Step 14. End.

Example 3.10 Following Example 3.9, the RF is applied to the test data of this category. Taking job #100 as an example. The decision rules applicable to this job are summarized in Table 3.5. The aggregation process is detailed in Table 3.6. Finally, the rules after reorganization are

$$\text{"If } x_{j2} \geq 1316 \text{ Then } \hat{C}_j = 1446.22\text{"}$$
$$\text{"If } x_{j2} \geq 1322 \text{ Then add } \hat{C}_j \text{ by } 2.27\text{"}$$
$$\text{"If } x_{j4} \geq 651.5 \text{ Then subtract } \hat{C}_j \text{ by } 7.00\text{"}$$

The first rule tells the user that the cycle time forecast of the job is around 1446.22 (hr). This rule is the most important (and accurate) rule. The user can stop reading if he/she doesn't want to continue reading. The user reads the rest of the rules

Table 3.5 Decision rules applicable to job #100

#	Rule content
1	"If $x_{j4} \geq 651.5$ Then $\hat{C}_j = 1425.22$"
2	"If $x_{j2} \geq 1322$ Then $\hat{C}_j = 1453.03$"
3	"If $x_{j2} \geq 1316$ Then $\hat{C}_j = 1446.22$"

Table 3.6 Aggregation process

Step #	Result
1	$\hat{C}_j = 1441.49$
2	$q = 1$
3	Rule #3 is chosen
4	RFII rule #1 is "If $x_{j2} \geq 1316$ Then $\hat{C}_j = 1446.22$"
5	$\hat{C}_j(\text{current}) = 1446.22$
6	Rule #3 is removed
7	$q = 2$
8	$q \ngtr 3$; go to Step 9
9	Rule #2 is chosen
10	$\Delta\hat{C}_j(\text{current}) = 2.27$
11	The premise of RFII rule #2 is "$x_{j2} \geq 1322$"
12	The consequence of RFII rule #2 is "add $\hat{C}_j$ by 2.269"
13	Return to Step 6
6	Rule #2 is removed
7	$q = 3$
8	$q \ngtr 3$; go to Step 9
9	Rule #1 is chosen
10	$\Delta\hat{C}_j(\text{current}) = -7.00$
11	The premise of RFII rule #3 is "$x_{j4} \geq 651.5$"
12	The consequence of RFII rule #3 is "add $\hat{C}_j$ by $-$ 7.00"
13	Return to Step 6
6	Rule #1 is removed
7	$q = 4$
8	$q > 3$; go to Step 14
14	End

from top to bottom, revising the predicted value successively, and finally learns that the predicted value is 1441.49 (hr).

Example 3.11 In Example 3.10, the performances of various explanation methods are summarized in Table 3.7. Both CART and RFII maximize the understandability of the prediction results. However, the approximation accuracy using CART is low, hindering its explainability. Furthermore, RFII achieves the highest explainability among highly understandable methods when only one rule is referenced.

Table 3.7 Explanation performances of various methods

Explanation method	Number of rules	Understandability	Explainability (RMSE) (hr)	Explainability using one rule (RMSE) (hr)
Black box	1	Very low	Not explained	Not explained
CART	1	Very high	98	98
RF	3	High	75.5	107
RFII	3	Very high	75.5	86

3.4 Domain Knowledge Application in Cycle Time Prediction Modeling

A cycle time prediction mechanism may not be consistent with domain knowledge [29, 33]. Domain knowledge is the domain expertise accumulated through continuous participation in a domain [34]. In the case of cycle time prediction, production planning and control engineers continuously predict the cycle time of each job in the factory. To do this, they monitor the factors that affect cycle time and interpret the relationship between these factors and cycle time, thereby forming their specific domain knowledge. For example, engineers may consider factory utilization as the factor that has the greatest impact on cycle time. However, the trained cycle time prediction model may not include this factor. Other considerations include

- The trained cycle time prediction model may contradict the domain knowledge. For example, the trained model may indicate that the higher the factory utilization, the shorter the expected cycle time, which contradicts domain knowledge.
- After extracting decision rules from the trained cycle time forecasting model, the order in which factors appear in the decision tree reflects the relative importance of these factors. Such information may be lost during the extraction of decision rules.
- The relative importance of factors revealed in the decision tree may not be consistent with domain knowledge.

3.4.1 Rule Modeling of Domain Knowledge

To confirm whether the trained cycle time prediction model is consistent to the domain knowledge, the domain knowledge must be expressed using the same XAI format as that used to express the cycle time prediction model (see Fig. 3.34).

In RFII, a decision rule has the following format:

$$\text{“If } x_{j1} \wedge_{k1} r_{k1} \text{ and} \ldots \text{and } x_{jP} \wedge_{kP} r_{kP} \text{ Then } \hat{C}_j = s_k\text{”} \quad (3.12)$$

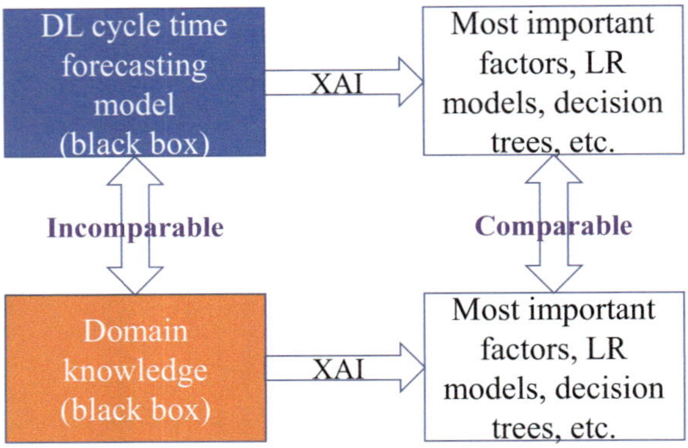

Fig. 3.34 Confirming whether the trained cycle time prediction model is consistent to the domain knowledge

where x_{jp} is the p-th attribute of job j; $p = 1 \sim P$; $j = 1 \sim n$ (the number of jobs); r_{kp} is the threshold of x_{jp} in the k-th decision rule; $k = 1 \sim K$ (the number of rules); $\wedge_{kp} \in \{<, \geq\}$ is the used logical operator; $\hat{C}_j$ is the cycle time forecast of job j; and s_k is the approximation result using decision rule k.

The following three domain knowledge are modeled (see Fig. 3.35):

- The most important factors affecting $\hat{C}_j$ include $x_{jp_1} \sim x_{jp_G}$: This can be confirmed if a decision tree (or rule) contains all these factors.
- x_{jp_1} is more important than x_{jp_2} to $\hat{C}_j$: This can be confirmed if x_{jp_1} appears much earlier (higher) than x_{jp_2} in the decision tree. Since a factor may appear multiple times in a decision tree, only the first occurrence is considered for comparison.
- $\hat{C}_j$ increases with x_{jp}: This can be guaranteed by comparing two decision rules, e.g., decision rules k_1 and k_2, with nearly identical conditions except for the m-th condition:

$$r_{k_1p} = r_{k_2p} \quad \text{and} \quad \wedge_{k_1p} = \wedge_{k_2p} \; \forall \, p \neq m \tag{3.13}$$

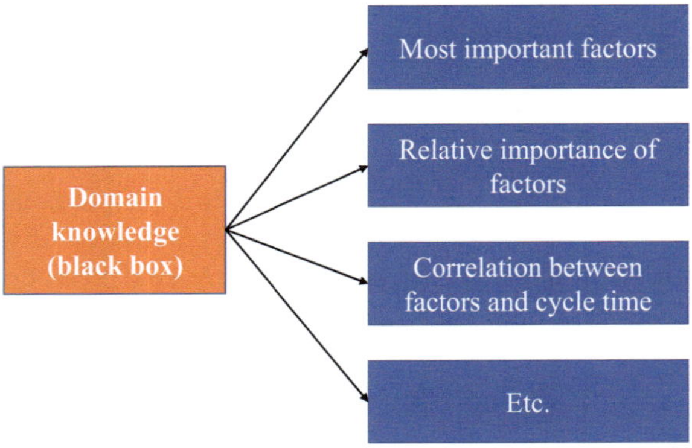

Fig. 3.35 Various types of domain knowledge to be compared

The two decision rules conform to the domain knowledge if either of the following conditions is true:

Condition I: If $\wedge_{k_1 m} = \wedge_{k_2 m}$ then $\begin{cases} s_{k_1} \geq s_{k_2} & \text{if} \quad r_{k_1 m} \geq r_{k_2 m} \\ s_{k_1} < s_{k_2} & \text{otherwise} \end{cases}$

Condition II: If $\wedge_{k_1 m} \neq \wedge_{k_2 m}$ then $\wedge_{k_1 m} r_{k_1 m} \cap \wedge_{k_2 m} r_{k_2 m} \neq \emptyset$.

Example 3.12 Job size is the number of workpieces processed simultaneously and is an integer. Factory utilization is determined by the percentage of time the busiest machine is in use and is in the range [0, 1]. Cycle time forecasts are usually measured in hours and can amount to thousands of hours. The following two decision rules match the domain knowledge "cycle time increases with factory utilization":

Rule #1: "If job size < 20 and factory utilization $\geq$ 85% then the estimated cycle time = 1305."

Rule #2: "If job size < 20 and factory utilization $\geq$ 90% then the estimated cycle time = 1420," where $\wedge_{k_1 m} = \wedge_{k_2 m} =$ "$\geq$", $r_{k_1 m} = 85\% < r_{k_2 m} = 90\%$, and $s_{k_1} = 1305 < s_{k_2} = 1420$. Rules #1 and #2 satisfy Condition I.

Example 3.13 The following two decision rules satisfy the second condition:

Rule #3: "If job size $\geq$ 22 and factory utilization < 95% then the estimated cycle time = 1527."

Rule #4: "If job size $\geq$ 22 and factory utilization $\geq$ 92% then the estimated cycle time = 1359," where $\wedge_{k_1 m} =$ "<" $\neq \wedge_{k_2 m} =$ "$\geq$" and $\wedge_{k_1 m} r_{k_1 m} =$ "<95%" $\cap \wedge_{k_2 m} r_{k_2 m}$ = "$\geq$ 92%" $\neq \emptyset$, all other conditions being the same.

In the third domain knowledge, two decision rules need to be compared, which means that

- The conformities of decision rules do not exist independently.
- Decision rules applicable to the same job need to be compared in pairs. In addition, decision rules extracted from an ANN (or DNN) using CART or the quick unbiased efficient statistical tree (QUEST) algorithm seem inconsistent. Such pairwise comparisons should be performed.

In addition, other types of domain knowledge need to be considered. However, because there may be many types of domain knowledge, experienced engineers in the factory may lack consensus on some types of domain knowledge.

Example 3.13 Three types of domain knowledge and associated checkpoints are given by a production manager in a DRAM factory, which are summarized in Table 3.8.

3.4.2 *Conformity Evaluation*

The conformity of decision rules to domain knowledge can be evaluated as follows. First, a score is assigned to each type of domain knowledge. Assume that the score of

Table 3.8 Three domain knowledge

No	Domain knowledge	Checkpoints
1	The most important factors affecting the approximated cycle time ($\hat{C}_j$) are job size (x_{j1}), WIP (x_{j2}) and factory utilization (x_{j6})	• Each decision tree contains the three factors • Each decision rule contains at least two of these factors
2	WIP (x_{j2}) is more important than factory utilization (x_{j6}) to approximating cycle time ($\hat{C}_j$)	• WIP (x_{j2}) appears before factory utilization (x_{j6}) in each decision tree
3	The approximated cycle time ($\hat{C}_j$) increases with each factor	• Two decision rules involve the same factors • The conditions of $T-1$ factors in the two rules are exactly the same. The conditions of only one factor in the two rules differ • If the value of a condition implies a higher value, the approximated cycle time ($\hat{C}_j$) should be longer, and vice versa

conforming to domain knowledge d is denoted as β_d; $d = 1 \sim D$. Then, the (overall) conformity of decision rule k to domain knowledge can be evaluated by the following formula:

$$\xi_k = \frac{\sum_{d=1}^{D} I_{kd}}{\sum_{d=1}^{D} \beta_d} \tag{3.14}$$

where

$$I_{kd} = \begin{cases} \beta_d & \text{if} \quad \text{rule } k \text{ meets domain knowledge } d \\ 0 & \text{otherwise} \end{cases} \tag{3.15}$$

Decision rules that do not conform to domain knowledge will be excluded from interpreting the cycle time prediction mechanism results using ANNs (or DNNs), as described below.

Example 3.14 In Example 3.13, job #1 is used as an example to illustrate the evaluation of the conformity of decision rules to domain knowledge in the following. Decision rules applicable to this job are extracted from the fitted RF and summarized in Table 3.9. The conformities of these decision rules to the three types of domain knowledge are also evaluated and shown in the same table. The production manager rates the importance of the three types of domain knowledge as 5, 3, and 4, respectively. On this basis, the overall conformity of each decision rule is assessed.

Table 3.9 Decision rules applicable to job #1

Decision rule no	Decision rule	Conforming to first domain knowledge	Conforming to second domain knowledge	Conforming to third domain knowledge	Overall conformity
1	If xj2 > = 1299 And xj6 > 0.858026 then Cj = 1316.5	V	X	V	0.75
2	If 1297 < = xj2 < 1341 and xj4 < = 809.5 and xj5 > = 390.169 and xj6 > = 0.837903 then Cj = 1321.6	V	X	V	0.75
3	If xj2 > = 1274.5 and 389.68 < = xj5 < 429.19 then Cj = 1270	X	V	V	0.58
4	If xj1 > = 21.5 and xj2 > = 1311.5 and xj4 > = 720 and xj5 > = 344.958 and xj6 > = 0.828446 then Cj = 1317.6667	V	X	V	0.75
5	If 160.5 < = xj3 < 167.5 and xj4 > = 709.5 and × 6 < 0.897438 then Cj = 1215.125	X	X	V	0.33
6	If xj2 > = 1311.5 and xj4 > = 709.5 and xj5 > = 365.567 and xj6 < 0.873495 then Cj = 1286	V	V	V	1
7	If xj1 < 23.5 and 1310.5 < = xj2 < 1324.5 And xj4 > = 698 then Cj = 1298.2222	V	V	V	1
8	If 1313.5 < = xj2 < 1332.5 and xj6 < 0.859609 then Cj = 1254.875	V	V	V	1
9	If 1307 < = xj2 < 1346.5 and xj4 > = 719 and xj5 < 424.755 then Cj = 1240.6	X	V	V	0.58
10	If xj2 < 1371 and 164.5 < = xj3 < 169.5 and 737 < = xj4 < 809 and xj5 > = 129.206 then Cj = 1308.4	X	V	V	0.58

3.4.3 Considering the Conformance to Domain Knowledge in Interpreting Cycle Time Prediction Process and Results

The RF used to explain the cycle time prediction process and results using ANNs (or DNNs) consists of many decision trees. From each decision tree, a decision rule can be applied to approximate the cycle time forecast of a job. Therefore, multiple rules can be applied simultaneously. In order to avoid user confusion during the interpretation process and make the interpretation consistent with domain knowledge, the improved RFII method reorganizes the decision rules extracted from the RF as follows:

Step 1. Extract the decision rules applicable to job j from all decision trees.
Step 2. Evaluate the conformities of these decision rules to domain knowledge.
Step 3. Exclude decision rules that have too low conformities to domain knowledge: The number of remaining decision rules is T.
Step 4. Average the outcomes of the remaining decision rules as $\hat{C}_j$.
Step 5. Set $q = 1$.
Step 6. Select the decision rule in which outcome is closest to $\hat{C}_j$.
Step 7. Let the q-th modified RFII rule be the decision rule.
Step 8. Replace the outcome of the decision rule with $\hat{C}_j$(current).
Step 9. Delete the decision rule.
Step 10. Increase q by 1.
Step 11. If $q > T$, go to Step 17; otherwise, go to Step 12.
Step 12. Find the decision rule in which outcome is closest to $\hat{C}_j$.
Step 13. Subtract $\hat{C}_j$(current) from the outcome of the decision rule, and divide the result by T: The result is denoted by $\Delta\hat{C}_j$(current).
Step 14. Replace the antecedent of the q-th modified RFII rule with that of the decision rule.
Step 15. Set the outcome of the q-th modified RFII rule to "add $\Delta\hat{C}_j$(current) to $\hat{C}_j$" or "subtract $-\Delta\hat{C}_j$(current) from $\hat{C}_j$."
Step 16. Return to Step 9.
Step 17. End.

Example 3.15 In Example 3.14, after removing the decision rules with low overall conformities, only six decision rules were left: decision rules #1, #2, #4, and #6–#8. $T = 6$. The remaining steps are detailed in Table 3.10. The final decision rules after aggregation are

Modified RFII decision rule #1: "If $x_{j1} < 23.5$ and $1310.5 \leq x_{j2} < 1324.5$ and $x_{j4} \geq 698$ then $\hat{C}_j = 1298.22$."

Modified RFII decision rule #2: "If $x_{j2} \geq 1311.5$ and $x_{j4} \geq 709.5$ and $x_{j5} \geq 365.57$ and $x_{j6} < 0.873$ then subtract 2.04 from $\hat{C}_j$."

Modified RFII decision rule #3: "If $x_{j2} \geq 1299$ and $x_{j6} < 0.858$ then add 2.97 to $\hat{C}_j$."

Modified RFII decision rule #4: "If $x_{j1} \geq 21.5$ and $x_{j2} \geq 1311.5$ and $x_{j4} \geq 720$ and $x_{j5} \geq 344.96$ and $x_{j6} \geq 0.828$ then add 3.24 to $\hat{C}_j$."

Table 3.10 Aggregation process using improved RFII

Step #	Result
4	$\hat{o}_j = 1299$
5	$q = 1$
6	Decision rule #7 is chosen
7	Modified RFII rule #1 is "If $x_{j1} < 23.5$ and $1310.5 \leq x_{j2} < 1324.5$ and $x_{j4} \geq 698$ then $\hat{o}_j = 1298.22$"
8	$\hat{o}_j(current) = 1298.22$
9	Decision rule #7 is removed
10	$q = 2$
11	$q \not> 6$; go to Step 12
12	Decision rule #6 is chosen
13	$\Delta\hat{o}_j(\text{current}) = -2.04$
14	The antecedent of modified RFII decision rule #2 is "If $x_{j2} \geq 1311.5$ and $x_{j4} \geq 709.5$ and $x_{j5} \geq 365.57$ and $x_{j6} < 0.873$"
15	The outcome of modified RFII rule #2 is "subtract 2.04 from $\hat{o}_j$"
16	Return to Step 9
9	Decision rule #6 is removed
10	$q = 3$
11	$q \not> 6$; go to Step 12
12	Decision rule #1 is chosen
13	$\Delta\hat{o}_j(\text{current}) = 2.97$
14	The antecedent of modified RFII decision rule #3 is "If $x_{j2} \geq 1299$ and $x_{j6} < 0.858$"
15	The outcome of modified RFII decision rule #3 is "add 2.97 to $\hat{o}_j$"
16	Return to Step 9
9	Decision rule #1 is removed
10	$q = 4$
11	$q \not> 6$; go to Step 12
12	Decision rule #4 is chosen
13	$\Delta\hat{o}_j(\text{current}) = 3.24$
14	The antecedent of modified RFII decision rule #4 is "If $x_{j1} \geq 21.5$ and $x_{j2} \geq 1311.5$ and $x_{j4} \geq 720$ and $x_{j5} \geq 344.96$ and $x_{j6} \geq 0.828$"
15	The outcome of modified RFII decision rule #4 is "add 3.24 to $\hat{o}_j$"
16	Return to Step 9
9	Decision rule #4 is removed
10	$q = 5$
11	$q \not> 6$; go to Step 12
12	Decision rule #2 is chosen

(continued)

Table 3.10 (continued)

Step #	Result
13	$\Delta\hat{o}_j(\text{current}) = 3.90$
14	The antecedent of modified RFII decision rule #5 is "If $1297 \le x_{j2} < 1341$ and $x_{j4} < 809.5$ and $x_{j5} \ge 390.169$ and $x_{j6} \ge 0.838$"
15	The outcome of modified RFII decision rule #5 is "add 3.90 to $\hat{o}_j$"
16	Return to Step 9
9	Decision rule #2 is removed
10	$q = 6$
11	$q \not> 6$; go to Step 12
12	Decision rule #8 is chosen
13	$\Delta\hat{o}_j(\text{current}) = -7.22$
14	The antecedent of modified RFII decision rule #6 is "If $1313.5 \le x_{j2} < 1332.5$ and $x_{j6} < 0.860$"
15	The outcome of modified RFII decision rule #6 is "subtract 7.22 from $\hat{o}_j$"
16	Return to Step 9
9	Decision rule #8 is removed
10	$q = 7$
11	$q > 6$; go to Step 17
17	End

Modified RFII decision rule #5: "If $1297 \le x_{j2} < 1341$ and $x_{j4} < 809.5$ and $x_{j5} \ge 390.169$ and $x_{j6} \ge 0.838$ then add 3.90 to $\hat{C}_j$."

Modified RFII decision rule #6: "If $1313.5 \le x_{j2} < 1332.5$ and $x_{j6} < 0.860$ then subtract 7.22 from $\hat{C}_j$."

The first modified RFII decision rule tells the user that the predicted value is around 1298.22 (hrs). This modified RFII decision rule is the most important decision rule. If the user does not want to continue reading, he/she can stop. The user reads the rest of the rules from top to bottom, corrects the predicted value gradually, and finally knows that the predicted value is 1299 (hrs).

References

1. T.C.T. Chen, Cycle time prediction and output projection, in *Production Planning and Control in Semiconductor Manufacturing: Big Data Analytics and Industry 4.0 Applications* (2022), pp. 41–62
2. D. Seth, N. Seth, P. Dhariwal, Application of value stream mapping (VSM) for lean and cycle time reduction in complex production environments: a case study. Prod. Plan. Control **28**(5), 398–419 (2017)

3. T. Chen, Y.-C. Wang, Fuzzy dynamic-prioritization agent-based system for job cycle time forecasting in a wafer fabrication plant. Complex Intell. Syst. **7**, 2141–2154 (2021)
4. Y.-C. Wang, T.-C.T. Chen, M.-C. Chiu, An explainable deep-learning approach for job cycle time prediction. Decis. Analytics **6**, 100153 (2023)
5. T.-C.T. Chen, Explainable artificial intelligence (XAI) in manufacturing, in *Explainable Artificial Intelligence (XAI) in Manufacturing: Methodology, Tools, and Applications* (2023), pp. 1–11
6. Y.C. Wang, T. Chen, Adapted techniques of explainable artificial intelligence for explaining genetic algorithms on the example of job scheduling. Expert Syst. Appl. **237**(A), 121369 (2024)
7. T.C.T. Chen, Explainable artificial intelligence (XAI) with applications, in *Explainable Ambient Intelligence (XAmI) Explainable Artificial Intelligence Applications in Smart Life* (2024), pp. 23–38
8. Y. Meng, N. Yang, Z. Qian, G. Zhang, What makes an online review more helpful: an interpretation framework using XGBoost and SHAP values. J. Theor. Appl. Electron. Commer. Res. **16**(3), 466–490 (2020)
9. Y.C. Wang, T.C.T. Chen, M.C. Chiu, An improved explainable artificial intelligence tool in healthcare for hospital recommendation. Healthc. Analytics **3**, 100147 (2023)
10. T.C.T. Chen, Explainable artificial intelligence (XAI), in *Explainable and Customizable Job Sequencing and Scheduling: Advancing Production Control and Management with XAI* (2025), pp. 1–20
11. T. Chen, Y.C. Wang, A two-stage explainable artificial intelligence approach for classification-based job cycle time prediction. Int. J. Adv. Manuf. Technol. **123**, 2031–2042 (2022)
12. H.C. Wu, T. Chen, CART–BPN approach for estimating cycle time in wafer fabrication. J. Ambient. Intell. Humaniz. Comput. **6**(1), 57–67 (2015)
13. T. Chen, Y.-C. Wang, H.-R. Tsai, Lot cycle time prediction in a ramping-up semiconductor manufacturing factory with a SOM-FBPN-ensemble approach with multiple buckets and partial normalization. Int. J. Adv. Manuf. Technol. **42**(11–12), 1206–1216 (2009)
14. T. Chen, H.-C. Wu, Y.-C. Wang, Fuzzy-neural approaches with example post-classification for estimating job cycle time in a wafer fab. Appl. Soft Comput. **9**, 1225–1231 (2009)
15. ConvNetJS, ConvnetJS demo: toy 2d classification with 2-layer neural network (2022). https://cs.stanford.edu/people/karpathy/convnetjs/demo/classify2d.html
16. P. Bholowalia, A. Kumar, EBK-means: a clustering technique based on elbow method and k-means in WSN. Int. J. Comput. Appl. **105**(9), 17–24 (2014)
17. T.C.T. Chen, C.W. Lin, Y.C. Wang, Visualization XAI techniques for facilitating schedulers' comprehension and interpretation of GA applications in job scheduling. Int. J. Hum.–Comput. Interact. 1–15 (2025)
18. I.N. Da Silva, D.H. Spatti, R.A. Flauzino, L.H.B. Liboni, S.F. dos Reis Alves, *Artificial Neural Networks* (Springer International Publishing, 2017)
19. GitHub, tensorflow (2022). https://github.com/tensorflow
20. S. Mantri, K. Bapat, Neural network based face recognition using MATLAB. Int. J. Comput. Sci. Eng. Technol. **1**(1), 6–9 (2011)
21. M. Green, U. Ekelund, L. Edenbrandt, J. Björk, J.L. Forberg, M. Ohlsson, Exploring new possibilities for case-based explanation of artificial neural network ensembles. Neural Netw. **22**(1), 75–81 (2009)
22. D. Gunning, M. Stefik, J. Choi, T. Miller, S. Stumpf, G.Z. Yang, XAI—explainable artificial intelligence. Sci. Robot. **4**(37), eaay7120 (2019)
23. E.M. Kenny, M.T. Keane, Twin-systems to explain artificial neural networks using case-based reasoning: comparative tests of feature-weighting methods in ANN-CBR twins for XAI, in *Twenty-Eighth International Joint Conferences on Artificial Intelligence* (2019), pp. 2708–2715
24. M.T. Ribeiro, S. Singh, C. Guestrin, Why should i trust you? Explaining the predictions of any classifier, in *Proceedings of the 22nd ACM SIGKDD International Conference on Knowledge Discovery and Data Mining* (2016), pp. 1135–1144
25. S. Na, L. Xumin, G. Yong, Research on k-means clustering algorithm: an improved k-means clustering algorithm, in *Third International Symposium on Intelligent Information Technology and Security Informatics* (2010), pp. 63–67

26. T. Chen, A job-classifying and data-mining approach for estimating job cycle time in a wafer fabrication factory. Int. J. Adv. Manuf. Technol. **62**(1), 317–328 (2012)
27. J. Liu, Q. Huang, C. Ulishney, C.E. Dumitrescu, Comparison of random forest and neural network in modeling the performance and emissions of a natural gas spark ignition engine. J. Energy Res. Technol. **144**(3), 032310 (2022)
28. T.C.T. Chen, C.W. Lin, A hybrid XAI-FCI approach for job cycle time range estimation: wafer fabrication as an example. Oper. Res. Int. J. **25**(2), 1–29 (2025)
29. T. Chen, Y.C. Wang, A modified random forest incremental interpretation method for explaining artificial and deep neural networks in cycle time prediction. Decis. Analytics J. **7**, 100226 (2023)
30. T. Chen, H.-C. Wu, M.-C. Chiu, A deep neural network with modified random forest incremental interpretation approach for diagnosing diabetes in smart healthcare. Appl. Soft Comput. **152**, 111183 (2024)
31. Y.C. Lin, T.C.T. Chen, Type-II fuzzy approach with explainable artificial intelligence for nature-based leisure travel destination selection amid the COVID-19 pandemic. Digit. Health **8**, 20552076221106320 (2022)
32. V. Rodriguez-Galiano, M. Sanchez-Castillo, M. Chica-Olmo, M.J.O.G.R. Chica-Rivas, Machine learning predictive models for mineral prospectivity: an evaluation of neural networks, random forest, regression trees and support vector machines. Ore Geol. Rev. **71**, 804–818 (2015)
33. T. Chen, Y.-C. Wang, Hybrid big data analytics and Industry 4.0 approach for projecting cycle time ranges. Int. J. Adv. Manuf. Technol. **120**, 279–295 (2022)
34. H. Yin, F. Fan, J. Zhang, H. Li, T.F. Lau, *The Importance of Domain Knowledge* (2020). https://blog.ml.cmu.edu/2020/08/31/1-domain-knowledge/

Chapter 4
Management Support

Abstract The results of cycle time prediction, analysis, and modeling can support various management activities, such as internal due date assignment, output projection, job sequencing and scheduling, competitiveness enhancement, customer relationship management, human resource management, and supply chain management. This chapter describes how to apply the results of cycle time prediction, analysis, and modeling to these management activities. Two management activities are especially focused on: internal due date assignment and output projection. Internal due date assignment refers to determining the date on which the manufacturing system can complete and deliver a potential order. The internal due date of a potential order can be determined by adding a certain allowance to the estimated completion time of the potential order. Advanced information technologies, such as cloud and edge computing, can also be applied to internal due date assignment. Therefore, a hybrid cloud and edge computing approach to enhance the effectiveness of internal due date assignment in advanced manufacturing systems is reviewed. In addition, imprecise output projection forecasts can lead to delivery delays, loss of potential orders, or unnecessary inventory accumulation, which is a critical challenge for manufacturing systems to overcome. Therefore, various machine learning or deep learning methods have been applied to output projection. Advanced information technologies, such as Industry 4.0 and explainable artificial intelligence, can also be applied to enhance the effectiveness and interpretability of output projection.

Keywords Cycle time prediction · Management support · Internal due date assignment · Output projection · Edge and cloud computing · Industry 4.0 · Explainable artificial intelligence (XAI)

T.-C. T. Chen, *Advances in Cycle Time Management*,
https://doi.org/10.1007/978-3-032-06911-5_4

4.1 Cycle Time Prediction, Analysis, and Modeling for Management Activities

The results of cycle time prediction, analysis, and modeling can support various management activities, such as internal due date assignment, output projection, job sequencing and scheduling, competitiveness enhancement, customer relationship management (CRM), human resource management, and supply chain management (SCM) [1] (see Fig. 4.1).

For example, Chen et al. [2] proposed a hybrid neural network and selective allowance method for internal due date assignment in wafer fabrication plants (wafer fabs). They first constructed a look-ahead self-organizing map (SOM)-fuzzy back propagation network (FBPN) to estimate the completion time of each job. Subsequently, a selective allowance policy was developed to determine the allowance to be added to the estimated job completion time. Compared with traditional methods, the look-ahead SOM-FBPN had many features (including incorporating future release

Fig. 4.1 Management activities supported by cycle time prediction, analysis, and modeling results

plans, classifying jobs, and incorporating expert opinions), among which the selective allowance policy was innovative in that it considered two different production control problems to avoid missing due dates.

Chen and Wu [3] proposed a fuzzy neural method for output prediction in semiconductor manufacturing plants. First, a fuzzy backpropagation network (FBPN) was constructed to generate fuzzy-valued completion time forecasts. It is worth noting that the upper bound of the fuzzy-valued completion time forecast contained the actual value at least for the training data. If the job was the last job of a possible order, the upper bound of its fuzzy-valued completion time forecast served as a basis for determining the internal due date. In addition, a fuzzy output projection function was proposed to project the output into each period.

Chen and Wang [4] proposed a nonlinear scheduling rule combined with a fuzzy neural remaining cycle time estimator to improve the scheduling performance in semiconductor manufacturing plants. The proposed scheduling rule was modified from the well-known fluctuation smoothing (FS) rule with three treatments. First, the forward-looking self-organizing map (SOM)-FBPN method was used to estimate the remaining cycle time of each job in the semiconductor manufacturing plant. Then, both the release time and the remaining cycle time of the job were normalized to balance their importance in the FS rule. Finally, the slack was derived by dividing the normalized release time by the normalized remaining cycle time. In this way, the proposed scheduling rule became a nonlinear rule.

For collaborative supply chain planning, supplier selection, and production and distribution planning, Che and Chiang [5] applied the rough set-based analytic hierarchy process (AHP) to establish an evaluation system for supplier procurement value. In addition, a cycle time estimation procedure was also established to effectively evaluate the operations in the collaborative supply chain. In addition, the assembly line balancing technique was used to ensure the smooth progress of production and distribution in the supply chain. Finally, a multi-objective optimization mathematical model considering procurement value, cost, cycle time, and the smoothness index was formulated and solved using a genetic algorithm (GA).

Some scheduling methods use cycle time forecasts as inputs [4]. Other scheduling methods try to minimize the average cycle time or cycle time variation [6, 7]. Still others have both characteristics [4] (see Fig. 4.2). These facts highlight the high relevance of cycle time prediction to job scheduling.

Cycle time management also impacts human resource management. The step cycle time of an operation is subject to uncertainty (or variation), and this local uncertainty may or may not affect the control of the cycle time. If it is a problem, workers with a stable work rhythm should be assigned to perform the operation. Otherwise, the selection of workers for this operation is more flexible. For example, in Fig. 4.3, the critical path contains operations that cannot be delayed: O11, O12, and O13 (i.e., operations in red). Therefore, skilled workers should be assigned to perform these operations, while new or unskilled workers should perform the remaining operations (i.e., operations in green) to allow for some delay but not so much as to delay the completion time of the product.

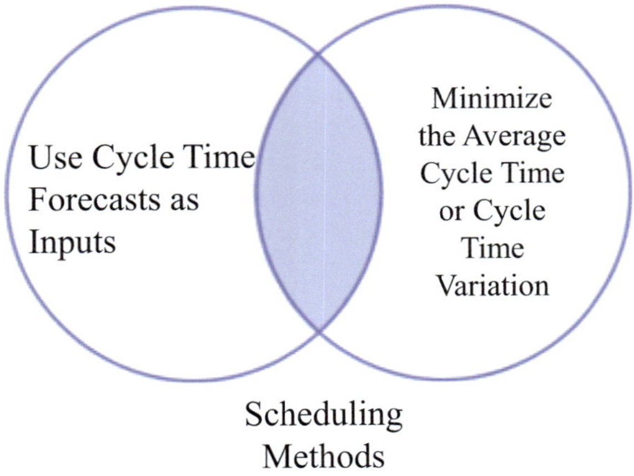

Fig. 4.2 High relevance of cycle time prediction to job scheduling

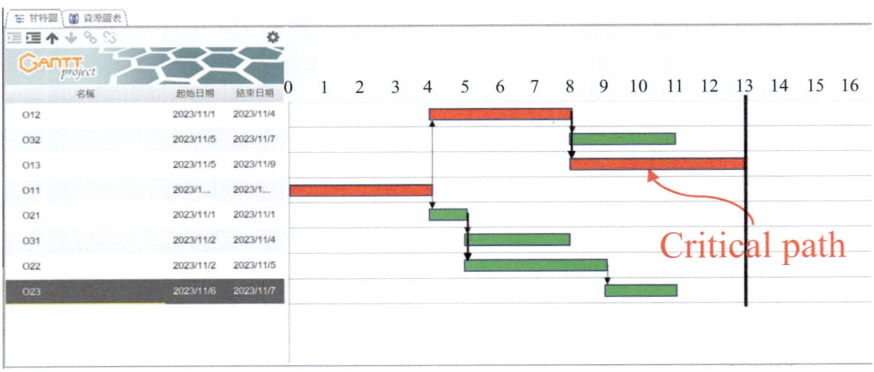

Fig. 4.3 Operations on the critical paths

In addition, reducing cycle time is considered a feasible way to enhance the competitiveness of manufacturing systems. In this regard, value stream mapping (VSM) is a lean manufacturing technique used to reduce cycle times. VSM aims to record, analyze, and improve the information and material flow in the manufacturing process [8], thereby eliminating various wastes and contributing to cycle time reduction.

In sum, according to Tan et al. [9], in computer component manufacturing, cycle time prediction results provide an important basis for scheduling control, material procurement, and due date assignment.

4.2 Internal Due Date Assignment

Internal due date assignment is one of the most common management activities supported by cycle time forecasting [10].

Due dates can be either internal or external. Internal due dates are usually based on when the production controller predicts the order will be completed, while external due dates are usually specified by customers and can be negotiated before being determined [11].

Internal due date assignment refers to the determination of the date by which the manufacturing system will complete and deliver a potential order [12, 13]. This is one of the most important CRM activities, and its results form the basis for the negotiated (external, actual) due date between the customer and the company. The closer the internal due date, the more attractive it is to the customer, as he/she can expect to receive the goods quickly. On the contrary, if the internal due date is further away, the company may have to reduce the price to compensate. Namely, price incentives are provided to entice customers to place orders.

Internal due date assignment is also an important topic in production control. Once the external due date for an order is determined, some production control measures may be taken, where appropriate, to ensure that the order is delivered on time. These conflicting considerations make internal due date assignment a challenging problem [14, 15].

4.3 Traditional Methods for Internal Due Date Assignment

Traditionally, the assignment of an internal due date follows the following steps (see Fig. 4.4):

- Virtually release the jobs of the possible order to the factory.
- Forecast the cycle time of each job in the possible order: Then, the completion time of the job can be estimated by adding the predicted cycle time to the release time.
- Estimate the completion time of the possible order: The completion time of the latest completed job determines the estimated completion time of the possible order.
- Add an allowance (including the overhead) to the estimated completion time to determine the internal due date.

4.3.1 *Virtually Releasing Jobs of the Possible Order*

Assuming that the monthly production capacity of the factory is C jobs and the number of working days per month is D, then the number of jobs that can be released

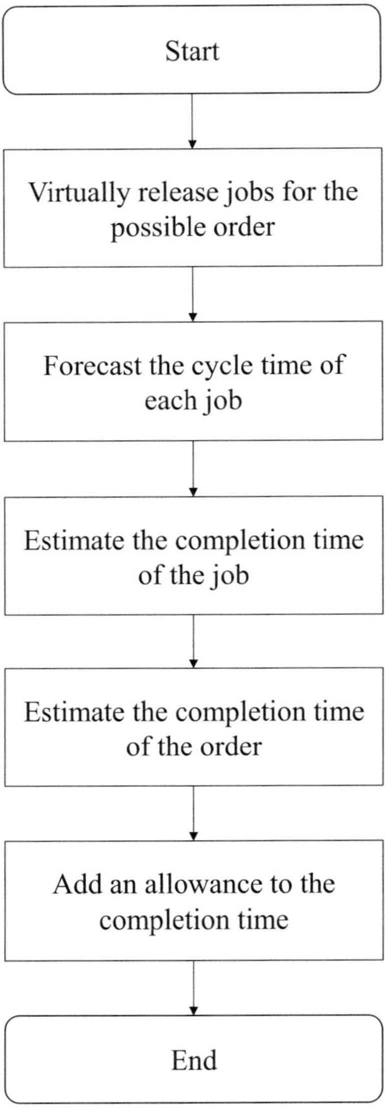

Fig. 4.4 Steps of internal due date assignment

into the factory every day is C/D, that is, the daily quota. Existing orders have already occupied some, and the jobs for a possible order may only occupy the remaining quota [16, 17] (see Fig. 4.5).

A possible order is decomposed into K jobs that will be virtually released to the factory within R days. Assume that the remaining quota on day t is ξ_t. Therefore, for this order, $\xi_t = 0 \; \forall t < R$. The release time of the k-th job in the possible order is determined as

Fig. 4.5 Determining the release times of jobs for a possible order [17]

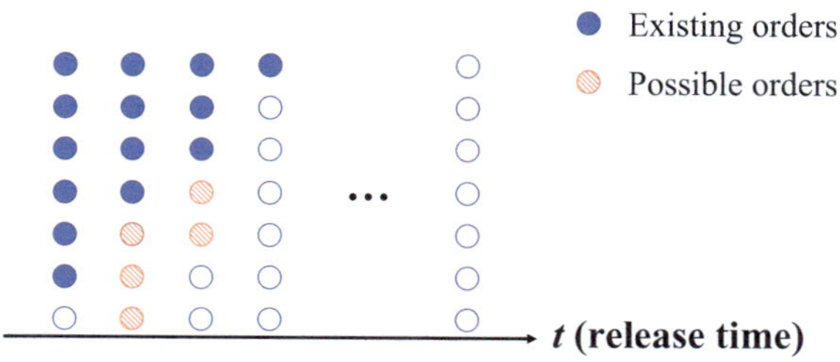

$$r_k = \begin{cases} \tau \text{ if } & \sum_{t=R}^{\tau-1} \xi_t < k \leq \sum_{t=R}^{\tau} \xi_t \\ \text{inf otherwise} & \end{cases} ; \forall k = 1 \sim K \tag{4.1}$$

4.3.2 *Estimating the Completion Time of Each Job in the Possible Order*

To estimate the completion time of a job, its cycle time needs to be predicted. Then the release time is added to the cycle time forecast to estimate the completion time of the job.

If cycle times are treated as a time series, the time series can be fitted to generate a series of cycle time forecasts, including those of jobs released in the future:

$$\hat{C}_j = f(\{C_l | l = 1 \sim j - 1\}) \tag{4.2}$$

$f()$ is the time series model for cycle time prediction. Otherwise, based on the job attributes and production conditions, the cycle time of job j, C_j, can be predicted using a machine learning (ML) or deep learning (DL) model as

$$\hat{C}_j = f\left(R_j, \boldsymbol{\alpha}_\mathbf{j}, \boldsymbol{\beta}_0\right) \tag{4.3}$$

where R_j is the release time of job j, $f()$ is the ML or DL model for cycle time prediction, $\boldsymbol{\alpha}_\mathbf{j}$ is the set of job (or order) attributes, and $\boldsymbol{\beta}_0$ is the set of production conditions in the factory. However, for a job that is released in the future, the production conditions of the factory are unknown and need to be estimated using time series analyses:

$$\hat{C}_k = f\left(R_k, \boldsymbol{\alpha}_\mathbf{k}, \hat{\boldsymbol{\beta}}_0\right); k = 1 \sim K \tag{4.4}$$

For example, Romanov et al. [18] applied type-II fuzzy logic to predict several time series data in an aircraft factory: employee work time fund (fluctuating time series), tool work time fund (fluctuating time series), performance ratio (growing time series), and area usage (growing time series). They also applied F-transformation to smooth these time series. Chen [19] proposed a fuzzy linear regression (FLR)-partial consensus fuzzy intersection (PCFI)-radial basis function (RBF) network method to predict the future work-in-process (WIP) level of a factory. In the FLR-PCFI-RBF method, a virtual expert committee was formed instead of convening many experts in the field. For each virtual expert, a corresponding FLR equation was constructed to predict the WIP level. Each FLR equation was fitted by solving two equivalent nonlinear programming (NLP) problems based on the opinions of the virtual experts. To aggregate these fuzzy WIP level forecasts, a two-step aggregation mechanism was used. First, PCFI was applied to aggregate the fuzzy WIP forecasts into a polygonal fuzzy number to improve the precision. Then, an RBF was constructed to defuzzify the polygonal fuzzy number and generate a representative/crisp value to improve the accuracy.

4.3.3 Estimating the Completion Time of the Possible Order

Based on the cycle time forecast of a job, the completion time of the job is estimated as

$$\hat{o}_k = R_k + \hat{C}_k \tag{4.5}$$

The last job in the possible order determines the order completion time

$$\hat{O} = \max_k \hat{o}_k \tag{4.6}$$

based on which the internal due date of the possible order can be assigned. Substituting (4.5) into (4.6) gives

$$\hat{O} = \max_k \left(R_k + \hat{C}_k\right) \tag{4.7}$$

Example 4.1 For a possible order, 10 jobs are released in sequence according to the remaining quota for each day (see Table 4.1). A time series is fitted to predict the cycle time of each job (see Fig. 4.6). After adding its release time, the completion time of the job is estimated, as summarized in Table 4.2. Then, the completion time of the potential order is determined by the completion time of the last job as the 161st day.

Table 4.1 Jobs released for a possible order

k	560	561	562	563	564	565	566	567	568	569
Release time (day)	93	94	94	95	95	95	95	95	96	96

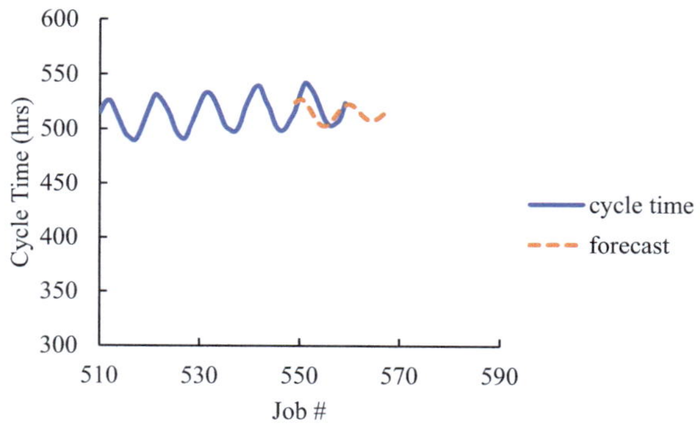

Fig. 4.6 Time series model for cycle time forecasting

Table 4.2 Completion time estimation results

k	560	561	562	563	564	565	566	567	568	569
Release time (day)	93	94	94	95	95	95	95	95	96	96
Cycle time forecast (hrs)*	523	520	515	511	508	509	512	516	519	521
Estimated completion time (day)	158	159	158	159	159	159	159	159	161	161

*: 8 working hours a day

Example 4.2 Ten jobs are released for a possible order, and the completion time of the possible order needs to be estimated. To this end, three factors are considered that affect the cycle time of each job (see Table 4.3). The values of the three factors for the 10 jobs have been estimated using time series analyses. The results are summarized in Table 4.4. Subsequently, a deep neural network (DNN), which is a feedforward neural network (FFNN) with multiple hidden layers [20] (see Fig. 4.7), is constructed to predict the cycle time of each job. The cycle time prediction results are summarized in Table 4.5. As a result, the estimated completion time of the possible order is $110 + 307/8 = 149$ (day).

Table 4.3 Cycle time-related data

j	R_j (day)	x_{j1} (%)	x_{j2}	x_{j3}	C_j (hrs)
1	69	68	207	50	272
2	69	64	258	46	267
3	69	66	228	62	276
…					
200	108	55	218	43	245

Table 4.4 Estimated values of the three factors for the 10 jobs

j	R_j (day)	x_{j1} (%)	x_{j2}	x_{j3}	C_j (hrs)
201	109	59	253	55	–
202	109	54	222	46	–
203	109	52	219	63	–
204	109	60	245	59	–
205	109	57	234	54	–
206	110	62	277	68	–
207	110	55	254	61	–
208	110	56	259	69	–
209	110	67	248	64	–
210	110	67	251	66	–

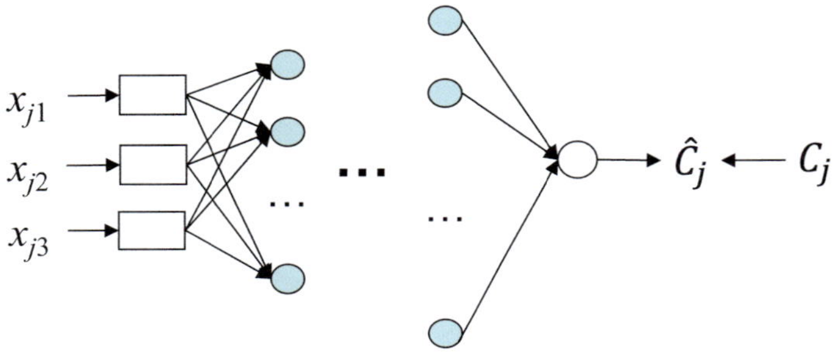

Fig. 4.7 DNN for cycle time forecasting

Table 4.5 Cycle time prediction results

j	R_j (day)	x_{j1} (%)	x_{j2}	x_{j3}	$\hat{C}_j$ (hrs)
201	109	59	253	55	258
202	109	54	222	46	259
203	109	52	219	63	245
204	109	60	245	59	269
205	109	57	234	54	259
206	110	62	277	68	306
207	110	55	254	61	277
208	110	56	259	69	260
209	110	67	248	64	276
210	110	67	251	66	307

Table 4.6 Due date setting rules

Method	Description
Constant flow allowance (CON) (or common due date)	• $D_j = \xi$ • D_j: due date of job j
Slack (SLK)	• $D_j = R_j + TP_j + \xi$ • TP_j: total processing time of job j • $\xi \in \mathbf{R}^+$ is the allowance (constant)
Total work (TWK)	• $D_j = \zeta \cdot TP_j$ • $\zeta \in \mathbf{R}^+$; $\zeta \geq 1$ is a constant
Processing plus wait (PPW)	• $D_j = \zeta \cdot TP_j + \xi$
Number of operations (NOP)	• $D_j = \zeta \cdot k_j$ • k_j is the number of operations of job j
Random allowance (RDW)	• $D_j = \varsigma$ where $p(C_j \geq \varsigma) \to 0$
Jobs in queue (JIQ)	• $D_j = \zeta \cdot \sum_k Q_k$ • Q_k: number of jobs in the queue before machine k
Jobs in system (JIS)	• $D_j = \zeta \cdot \text{WIP}$ • WIP: work-in-process level in the factory
Selective allowance policy [2]	• $D_j = \hat{C}_j + \xi_j$ • ξ_j: allowance of job j, which is proportional to the difficulty of accelerating the progress of this job

4.3.4 *Adding an Allowance to the Completion Time*

Cycle time forecasts rarely equal actual values. Therefore, if the cycle time forecasting method is not biased, the average probability of completing a job before the estimated completion time is only 50%, and the same is true for the possible order containing these jobs. Therefore, if the internal due date of a possible order is set to its completion time, it is likely to be missed.

To solve this problem, traditionally some methods add an allowance to the estimated completion time of the possible order to determine the internal due date, which is the so-called due date setting rule [21], as shown in Table 4.6.

4.3.5 *Estimating the Upper Bound of the Completion Time Using Fuzzy Methods*

Another approach is to estimate the upper bound of the completion time of each job in the possible order, and then determine the internal due date by the upper bound of the completion time of the last job. To this end, various fuzzy methods are applicable.

The simplest fuzzy method is a fuzzy linear regression (FLR), in which the cycle time of job j is predicted as

$$\tilde{C}_j = \tilde{w}_0(+)\sum_{p=1}^{P}\tilde{w}_p x_{jp} \tag{4.8}$$

where x_{jp} is the p-th attribute of job j; $p = 1 \sim P$; $j = 1 \sim n$; $\tilde{C}_j$ is the fuzzy cycle time forecast; and $\tilde{w}_p$ is the coefficient for x_{jp}. . (+) indicates fuzzy multiplication. When all fuzzy parameters are given in triangular fuzzy numbers (TFNs), the result is also a TFN:

$$\tilde{C}_j = \left(C_{j1}, C_{j2}, C_{j3}\right) \tag{4.9}$$

where

$$C_{j1} = w_{01} + \sum_{p=1}^{P} w_{p1}x_{jp} \tag{4.10}$$

$$C_{j2} = w_{02} + \sum_{p=1}^{P} w_{p2}x_{jp} \tag{4.11}$$

$$C_{j3} = w_{03} + \sum_{p=1}^{P} w_{p3}x_{jp} \tag{4.12}$$

The following LP model derives the values of fuzzy parameters by minimizing the average range of fuzzy forecasts [3, 6]:

(LP Model I)

$$\text{Min} \quad Z_1 = \sum_{j=1}^{n}\left(C_{j3} - C_{j1}\right) \tag{4.13}$$

subject to

$$C_j \geq (1-s)C_{j1} + sC_{j2}; j = 1 \sim n \tag{4.14}$$

$$C_j \leq (1-s)C_{j3} + sC_{j2}; j = 1 \sim n \tag{4.15}$$

$$C_{j1} = w_{01} + \sum_{p=1}^{P} w_{p1}x_{jp}; j = 1 \sim n \tag{4.16}$$

$$C_{j2} = w_{02} + \sum_{p=1}^{P} w_{p2} x_{jp};\ j = 1 \sim n \tag{4.17}$$

$$C_{j3} = w_{03} + \sum_{p=1}^{P} w_{p3} x_{jp};\ j = 1 \sim n \tag{4.18}$$

$$w_{p1} \le w_{p2} \le w_{p3};\ p = 0 \sim P \tag{4.19}$$

where C_j is the actual value and s is the pre-specified satisfaction level within [0, 1]. Constraints (4.14) and (4.15) ensure that the membership of the actual value in the fuzzy forecast is greater than s. A larger value of s results in a wider fuzzy forecast.

Example 4.3 Cycle time-related data have been collected for 24 jobs and are presented in Table 4.7, in which the cycle time of job j, C_j, is to be predicted based on the values of six job attributes, $x_{j1} \sim x_{j6}$, using an FLR equation. The data of the first 1/2 jobs are used as the training data, while the rest are left for evaluating the forecasting performance, $s = 0.3$. The MATLAB code for fitting the FLR equation is presented in Fig. 4.8. The fitted FLR equation is

$$\begin{aligned}\tilde{y}_j = {} & (0.488, 157.4, 157.4)(+)(-11.58, -11.58, -11.58)(\times)x_{j1} \\ & (+)(2.258, 2.258, 2.258)(\times)x_{j2}(+)(0.018, 0.018, 0.018)(\times)x_{j3} \\ & (+)(-2.188, -2.188, -2.188)(\times)x_{j4}(+) \\ & (-0.004, -0.004, -0.004)(\times)x_{j5}(+)(89.44, 89.44, 89.44)(\times)x_{j6}\end{aligned}$$

The forecasting results are summarized in Fig. 4.9.

Subsequently, the release time of a job is added to the fuzzy cycle time forecast of the job to estimate the fuzzy completion time:

$$\tilde{o}_j = R_j + \tilde{C}_j = \left(o_{j1}, o_{j2}, o_{j3}\right) \tag{4.20}$$

Then, the upper bound of the fuzzy completion time of the last job determines the internal due date.

Table 4.7 Cycle time-related data of 28 jobs

j	x_{j1}	x_{j2}	x_{j3}	x_{j4}	x_{j5}	x_{j6} (%)	C_j (hrs)
1	22	1320	165	748	421	85	1298
2	23	1261	169	734	198	81	1098
3	22	1187	159	635	287	93	1122
…							
24	23	1347	184	681	370	97	1547

```
min=Z11+Z12;
Z11=y13-y11+y23-y21+y33-y31+y43-y41+y53-y51+y63-y61;
Z12=y73-y71+y83-y81+y93-y91+y103-y101+y113-y111+y123-y121;
0.7*y11+0.3*y12<=1298;
0.7*y21+0.3*y22<=1098;
0.7*y31+0.3*y32<=1122;
...
0.7*y121+0.3*y122<=1066;
0.7*y13+0.3*y12>=1298;
0.7*y23+0.3*y22>=1098;
0.7*y33+0.3*y32>=1122;
...
0.7*y123+0.3*y122>=1066;
y11=w01+w11*22+w21*1320+w31*165+w41*748+w51*421+w61*0.85;
y21=w01+w11*23+w21*1261+w31*169+w41*734+w51*198+w61*0.81;
y31=w01+w11*22+w21*1187+w31*159+w41*635+w51*287+w61*0.93;
...
y121=w01+w11*23+w21*1289+w31*157+w41*815+w51*462+w61*0.78;
y12=w02+w12*22+w22*1320+w32*165+w42*748+w52*421+w62*0.85;
y22=w02+w12*23+w22*1261+w32*169+w42*734+w52*198+w62*0.81;
y32=w021+w12*22+w22*1187+w32*159+w42*635+w52*287+w62*0.93;
...
y122=w02+w12*23+w22*1289+w32*157+w42*815+w52*462+w62*0.78;
y13=w03+w13*22+w23*1320+w33*165+w43*748+w53*421+w63*0.85;
y23=w03+w13*23+w23*1261+w33*169+w43*734+w53*198+w63*0.81;
y33=w03+w13*22+w23*1187+w33*159+w43*635+w53*287+w63*0.93;
...
y123=w03+w13*23+w23*1289+w33*157+w43*815+w53*462+w63*0.78;
w01-w02<=0;
w02-w03<=0;
w11-w12<=0;
w12-w13<=0;
w21-w22<=0;
w22-w23<=0;
...
w61-w62<=0;
w62-w63<=0;
@free(w01); @free(w02); @free(w03); @free(w11); @free(w12); @free(w13); @free(w21);
@free(w22); @free(w23); @free(w31); @free(w32); @free(w33); @free(w41); @free(w42);
@free(w43); @free(w51); @free(w52); @free(w53); @free(w61); @free(w62); @free(w63);
```

Fig. 4.8 MATLAB code for fitting the FLR equation

Example 4.4 The last job in a possible order is released at time 472 (hrs) with the following attributes:

$$x_{j1} = 23; x_{j2} = 1269; x_{j3} = 158; x_{j4} = 721; x_{j5} = 326; x_{j6} = 85\%$$

Then, according to the fitted FLR equation, $\tilde{C}_j = (1100, 1256, 1256)$ (hrs). After adding the release time to $\tilde{C}_j$, $\tilde{o}_j = (1572, 1728, 1728)$ (hrs). Therefore, the upper bound of the last job is 1728 h, which is used to determine the internal due date of the possible order.

For the same purpose, Chen et al. [2] constructed a fuzzy feedforward neural network (FFFNN) with the following configuration:

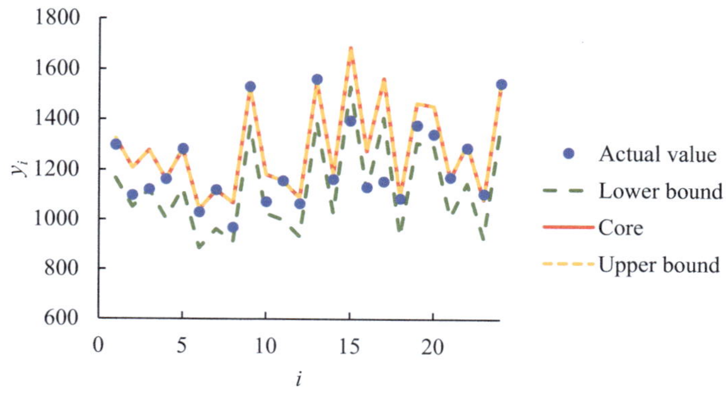

Fig. 4.9 Forecasting results using FLR

- Inputs: P inputs corresponding to the attributes of job j and the production conditions when the job is released into the factory. It is worth noting that for jobs released in the future, production conditions are unknown and must be estimated in advance.
- Single hidden layer: Generally one or two hidden layers are more beneficial for the convergence property of the FFFNN.
- Number of neurons in the hidden layer: The same as that in the input layer. Such a treatment has been adopted by many studies [22]. Two other formulas for determining the suitable number of neurons in the hidden layer are provided below:

  (i) Number of neurons in the hidden layer × = (number of neurons in the input layer + number of neurons in the output layer)/2.
  (ii) Number of neurons in the hidden layer × = (number of neurons in the input layer · number of neurons in the output layer)$^{1/2}$.

- Output: The fuzzy cycle time forecast (normalized value) ($o_j = N(\hat{C}_j)$) of the job.
- Actual value: The cycle time (normalized value) of the job:

$$a_j = N(C_j) = \frac{\left(C_j - \min_l C_l\right)}{(\max_l C_l - \min_l C_l)} \cdot 0.8 + 0.1 \tag{4.21}$$

In this way, $a_j \in [0.1, 0.9]$.

- Network learning rule: Fuzzified Levenberg–Marquardt (LM) algorithm.
- Transformation function: Log-Sigmoid function:

$$f(x) = \frac{1}{1 + e^{-x}} \tag{4.22}$$

- Learning rate (η): 0.01 ~ 1.0.
- Batch learning.

In addition, only the threshold of the node on the hidden layer θ^o is fuzzified.

Two phases are involved at the training stage. At first, in the forward phase, inputs are multiplied with weights, summated, and transferred to the hidden layer. Then, activated signals are outputted from the hidden layer as follows:

$$h_{jl} = \frac{1}{1 + e^{-n_{jl}^h}} \tag{4.23}$$

where

$$n_{jl}^h = I_{jl}^h - \theta_l^h \tag{4.24}$$

$$I_{jl}^h = \sum_p (w_{pl}^h x_{jp}) \tag{4.25}$$

h_j's are also transferred to the output layer with the same procedure. Finally, the output of the FFFPN is generated as follows:

$$o_j = \frac{1}{1 + e^{-n_j^o}} \tag{4.26}$$

where

$$n_j^o = I_j^o - \theta^o \tag{4.27}$$

$$I_j^o = \sum_l (w_l^o h_{jl}) \tag{4.28}$$

The procedure for determining the values of network parameters is composed of two stages. At the first stage, the LM algorithm is applied to optimize the values of network parameters. Subsequently, θ^o is fuzzified to generate a fuzzy network output that contains the actual value:

$$\mu_{\tilde{o}_j}(a_j) \geq 0 \tag{4.29}$$

where

$$\tilde{o}_j = \frac{1}{1 + e^{-\tilde{n}_j^o}} \tag{4.30}$$

$$\tilde{n}_j^o = I_j^{o*}(-)\tilde{\theta}^o \tag{4.31}$$

(−) indicates fuzzy subtraction. Equation (4.29) is equivalent to

$$o_{j1} \leq a_j \leq o_{j3} \tag{4.32}$$

where

$$o_{j1} = \frac{1}{1 + e^{-n_{j1}^o}} \tag{4.33}$$

$$o_{j2}^* = \frac{1}{1 + e^{-n_{j2}^{o*}}} \tag{4.34}$$

$$o_{j3} = \frac{1}{1 + e^{-n_{j3}^o}} \tag{4.35}$$

where

$$n_{j1}^o = I_j^{o*} - \theta_3^o \tag{4.36}$$

$$n_{j2}^{o*} = I_j^{o*} - \theta_2^{o*} \tag{4.37}$$

$$n_{j3}^o = I_j^{o*} - \theta_1^o \tag{4.38}$$

Substituting Eq. (4.37) into Eq. (4.34) gives

$$o_{j2}^* = \frac{1}{1 + e^{-(I_j^{o*} - \theta_2^{o*})}} \tag{4.39}$$

Therefore,

$$I_j^{o*} = -\ln\left(\frac{1}{o_{j2}^*} - 1\right) + \theta_2^{o*} \tag{4.40}$$

In addition, substituting Eqs. (4.36) and (4.38) into Constraint (4.32) gives

$$\frac{1}{1 + e^{-(I_j^{o*} - \theta_3^o)}} \leq a_j \tag{4.41}$$

$$\theta_3^o \geq \ln\left(\frac{1}{a_j} - 1\right) + I_j^{o*} \tag{4.42}$$

Substituting Eq. (4.40) into Constraint (4.42) gives

$$\theta_3^o \geq \ln\left(\frac{1}{a_j} - 1\right) - \ln\left(\frac{1}{o_{j2}^*} - 1\right) + \theta_2^{o*} \tag{4.43}$$

To minimize the cycle time range,

$$\theta_3^o = \max_j\left(\ln\left(\frac{1}{a_j} - 1\right) - \ln\left(\frac{1}{o_{j2}^*} - 1\right) + \theta_2^{o*}\right) \tag{4.44}$$

$$\theta_2^{o*} - \theta_3^o = -\max_j\left(\ln\left(\frac{1}{a_j} - 1\right) - \ln\left(\frac{1}{o_{j2}^*} - 1\right)\right) \tag{4.45}$$

Subsequently, after substituting Eq. (4.38) into Eq. (4.33), we have

$$o_{j1} = \frac{1}{1 + e^{-(I_j^{o*} - \theta_3^o)}} \tag{4.46}$$

$$o_{j1} = \frac{1}{1 + e^{-(I_j^{o*} - \theta_2^{o*} + \theta_2^{o*} - \theta_3^o)}} \tag{4.47}$$

According to Eq. (4.40),

$$I_j^{o*} - \theta_2^{o*} = -\ln\left(\frac{1}{o_{j2}^*} - 1\right) \tag{4.48}$$

Substituting Eqs. (4.48) and (4.45) into Eq. (4.47) gives

$$o_{j1} = \frac{1}{1 + e^{-\left(-\ln\left(\frac{1}{o_{j2}^*}-1\right)-\max_j\left(\ln\left(\frac{1}{a_j}-1\right)-\ln\left(\frac{1}{o_{j2}^*}-1\right)\right)\right)}} \tag{4.49}$$

Similarly

$$o_{j3} = \frac{1}{1 + e^{-\left(-\ln\left(\frac{1}{o_{j2}^*}-1\right)-\min_j\left(\ln\left(\frac{1}{a_j}-1\right)-\ln\left(\frac{1}{o_{j2}^*}-1\right)\right)\right)}} \tag{4.50}$$

After converting back to the unnormalized value,

$$U\left(o_{j3}\right) = \frac{\left(o_{j3} - 0.1\right)}{0.8} \cdot \left(\max_l C_l - \min_l C_l\right) + \min_l C_l \tag{4.51}$$

Example 4.5 To determine the internal due date of a possible order, the upper bounds on the cycle times of the ten jobs of the possible order are to be estimated. According to the training data (see Table 4.8),

Table 4.8 Training data

j	R_j (day)	a_j (normalized cycle time)	o^*_{j2} (FFNN output)
1	69	0.461	0.436
2	69	0.428	0.396
3	69	0.487	0.493
…			
200	108	0.290	0.299

$$\min_j\left(\ln\left(\frac{1}{a_j}-1\right)-\ln\left(\frac{1}{o^*_{j2}}-1\right)\right)=\min\left(\ln\left(\frac{1}{0.461}-1\right)\right.$$
$$\left.-\ln\left(\frac{1}{0.436}-1\right),\ldots,\ln\left(\frac{1}{0.290}-1\right)-\ln\left(\frac{1}{0.299}-1\right)\right)=-0.281.$$

Then, the upper bounds on the cycle times of the ten jobs of the possible order can be estimated. For example,

$$o_{201,3}=\frac{1}{1+e^{-\left(-\ln\left(\frac{1}{0.371}-1\right)-(-0.281)\right)}}=0.438$$

After converting back to the unnormalized value, the upper bound on the cycle time of this job is

$$\frac{(0.438-0.1)}{0.8}\cdot(340-216)+216=268\ (\text{hrs})$$

where 340 and 216 are the maximum and minimum cycle times in the training data, respectively. Finally, the internal due date of the possible order is determined as day 149 (Table 4.9).

4.4 Edge Computing Application in Internal Due Date Assignment

Edge computing has been used in manufacturing to perform real-time data processing closer to the source, thereby alleviating the limited accessibility and confidentiality risks of cloud applications. However, existing edge computing applications in manufacturing systems are mostly used for machine status monitoring (including predictive maintenance) [23, 24], material transportation tracking [25], resource management, process control, production planning (for machine tools) [26], quality inspection [27, 28], etc., rather than for internal due date assignment or other management activities.

Table 4.9 Ten jobs of the possible order

j	R_j (day)	o^*_{j2} (FFNN output)	o_{j3}	Cycle time upper bound (hrs)	Completion time upper bound (day)
201	109	0.371	0.438	268	143
202	109	0.376	0.444	269	143
203	109	0.287	0.348	254	141
204	109	0.441	0.511	280	144
205	109	0.379	0.447	270	143
206	110	0.679	0.737	315	149
207	110	0.492	0.562	288	146
208	110	0.388	0.456	271	144
209	110	0.487	0.557	287	146
210	110	0.685	0.743	316	149

Internal due date assignment may require DL applications in the cloud for job cycle time prediction, which is often difficult to interpret and/or communicate. In addition, such DL models need to be retrained after new data coming in, which may not be efficient in practice. To address these issues, edge computing is expected to simplify DL applications and enable sales engineers to fulfill internal due date assignment tasks at the edge.

4.4.1 System Architecture and Implementation Procedure

Chen and Lin [17] established a cloud and edge computing system to improve the effectiveness of internal due date assignment in advanced manufacturing systems. The system architecture of the cloud and edge computing system for internal due date assignment is shown in Fig. 4.10. They aimed to use a random forest (RF) to approximate the DNN for cycle time prediction and deploy RF rules that were easier to interpret, easier to communicate, and data more secure at the edge.

The implementation process of the cloud and edge computing system for internal due date assignment includes the following steps:

Step 1. (Industrial IoT sensors) Collect/recollect production data required for internal due date assignment: In some advanced manufacturing systems, many sensors and computing units have been embedded in machines to collect state variable identification (SVID) data such as machine and job status. Radio frequency identification (RFID) tags are also attached to containers to provide real-time information about the location and progress of jobs. Instruments, robots, and machine tools are other field devices that collect the required production data [23]. These industrial IoT sensors in the factory act as internal edge points to

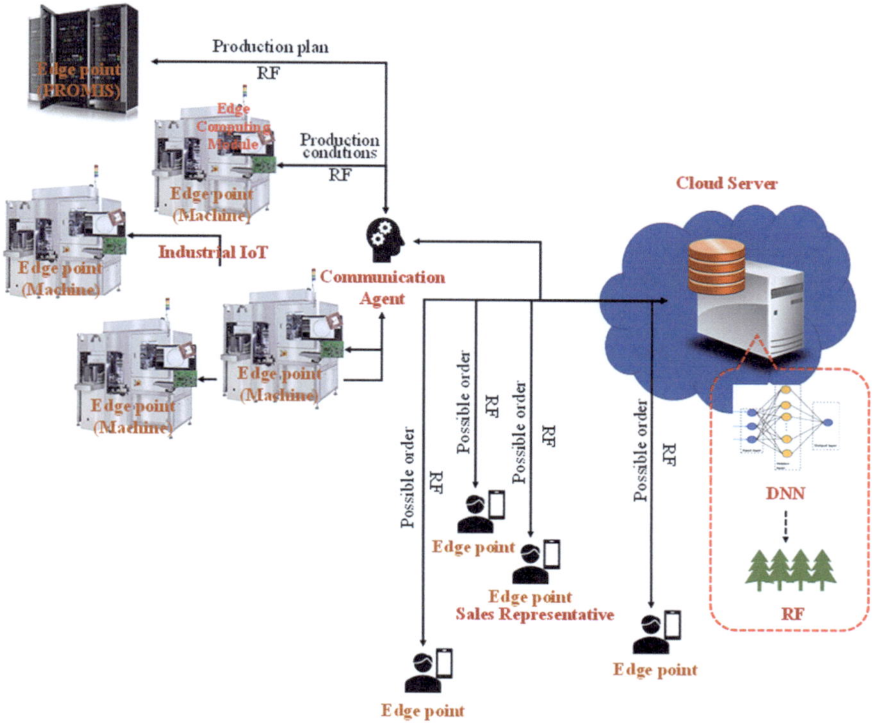

Fig. 4.10 System architecture of the cloud and edge computing system for internal due date assignment [17]

collect real-time status data of machines and jobs. The latest performances of machines and the entire factory can be retrieved from the production management information system (PROMIS). The daily remaining quota is based on the latest production plan. The cloud and edge computing system infrastructure for internal due date assignment allows edge points to achieve flexible communication using various communication protocols under standardized communication models, thereby uploading the collected production data to the cloud server for further analyses.

Step 2. (Communication agent) Upload production data to the cloud server for analysis using cloud computing.

Step 3. (Cloud server) Build/rebuild a DNN to estimate the completion time of potential orders.

Step 4. (Cloud server) Approximate the trained DNN using an RF for each edge point.

Step 5. (Sales representative) Applied the RF (as the edge computing application) to assign internal due dates for possible orders.

Figure 4.11 provides a flowchart to illustrate the process.

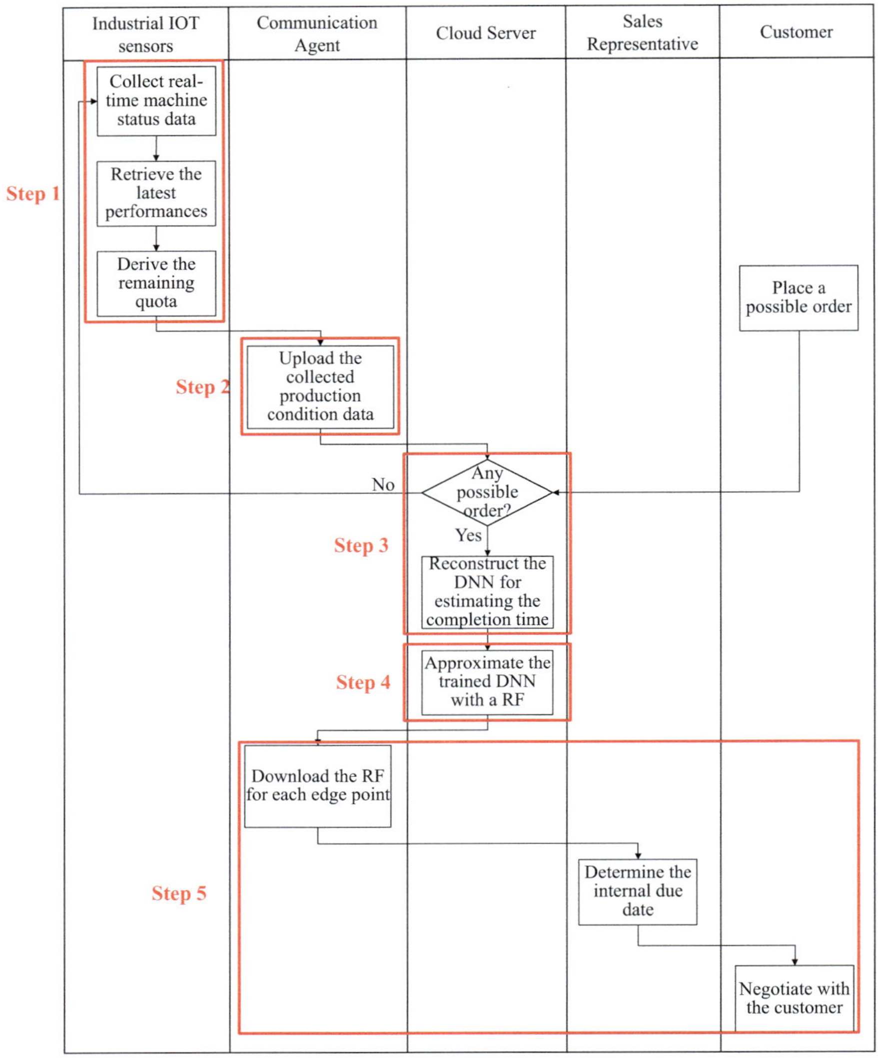

Fig. 4.11 Implementation procedure of the cloud and edge computing system for internal due date assignment [17]

The connection between cloud and edge computing, including their respective roles and interactions, is described as follows:

- Machines (internal edge points) collect real-time production data, and sales representative (external edge points) collect the contents of possible orders.
- The collected data is uploaded to the cloud server and used to reconstruct the DNN (i.e., the cloud computing application) for job completion time prediction. The cloud server also uses an RF for each edge point to approximate the DNN.

Table 4.10 Functions and algorithms of the edge, fog, and cloud layers [17]

Layer	Functions	Algorithm
Edge layer	• Data sensing • Protocol conversion • Real-time analysis • Real-time control • Real-time monitoring • Simulation	Edge computing
Fog layer	• Inventory management • Material distribution • Proactive maintenance, repair, and overhaul (MRO) • Resource management • Smart scheduling	Fog computing
Cloud layer	• Big data analysis • Cloud storage • Services collaboration • Services management • Supply–demand matching • Task management	Cloud computing

- The cloud server transmits the RF to each edge point, or each edge point downloads the RF from the cloud.
- Machines use their RFs to reinforce production control (i.e., internal edge computing applications), while sales representatives apply RFs for internal due date assignment (i.e., external edge computing applications).

Qi and Tao [29] believed that smart manufacturing systems should connect more devices to the Internet to collect more data at various stages of the product life cycle. Such data can then be analyzed by various applications and services in the cloud, thereby facilitating large-scale collaboration. They added a fog computing layer between the edge layer and the cloud layer, and applied fog computing, edge computing, and cloud computing technologies to implement the corresponding functions, as summarized in Table 4.10.

4.4.2 Edge Computing Application

If all sales representatives were connected to the cloud server and applied the trained DNN, the cloud server would be overloaded. In addition, all product attributes and production conditions would be at risk of being leaked. To address these issues, an RF is used to approximate the trained DNN for each sales representative to apply in internal due date assignment.

The RF contains B decision trees. Each decision tree considers only a subset of job attributes, which is called feature bagging [30], thus making the decision trees diverse. For a similar purpose, examples used to train the decision trees are also randomly selected by replacement, i.e., the concept of bootstrapping [31]. In this way, these decision trees provide a local approximation to the trained DNN following the idea of locally interpretable model-agnostic explanations (LIME).

For job k, only one decision rule applies in each decision tree with the following format [32–34]:

$$\text{"If } x_{j1} \wedge_{q1} (b) c_{q1}(b) \text{ and} \ldots \text{and } x_{jL} \wedge_{qL} (b) c_{qL}(b) \text{ then } \hat{C}_j(b) = \xi_j(b)\text{"} \quad (4.52)$$

where $c_{ql}(b)$ is the threshold for x_{jl} in the q-th decision rule of decision tree b; $x_{jl} \in \mathbf{r} \cup \boldsymbol{\alpha} \cap \boldsymbol{\beta}_0$. $q = 1 \sim Q(b)$ (the number of rules); $b = 1 \sim B$. $\wedge_{ql}(b) \in \{=, <, \geq, within\}$ is the used logical operator; $\hat{C}_j(b)$ is the approximated value of the estimated completion time of job k by decision tree b; and $\xi_j(b)$ is the outcome of decision tree b.

In contrast, using the DNN, all data are required and the original data have to be revealed:

$$\text{"If } x_{j1} = \ldots \text{ And } x_{j2} = \ldots \text{And } x_{jP} = \ldots \text{Then } \hat{C}_j = \ldots\text{"} \quad (4.53)$$

Some of the required data, such as the utilization of the bottleneck machine or the entire factory, should be kept confidential. Table 4.11 compares the suitability of the DNN and the RF for internal due date assignment. Therefore, $L \leq P$.

Each decision tree is trained to minimize the root mean square error (RMSE) of approximating $\hat{C}_j$ with $\xi_j(b)$:

$$\text{RMSE}(b) = \sqrt{\frac{\sum_{k=1}^{K} (\hat{C}_j - \xi_j(b))^2}{K}} \quad (4.54)$$

subject to constraints such as the minimum number of examples of a branch, the maximum number of nodes, and the tallest tree height/depth [35]. The process of building each decision tree goes through three phases: tree growth, stopping, and pruning [36]. The completion times of a job estimated by all decision trees are aggregated by averaging [37, 38]:

Table 4.11 Comparison of the DNN and the RF for internal due date assignment

	DNN	RF
Data completeness	All data required	Only partial data required (due to feature bagging)
Data security	Raw data revealed	Only the range of data revealed
Easiness to understand and communicate	Very low	Very high
Accuracy	Very high	High (if many decision trees are used)

$$\xi_j = \frac{\sum_{b=1}^{B} \xi_j(b)}{B} \tag{4.55}$$

Then, the completion time of the possible order is given by the maximum output time:

$$O = \max_j \xi_j \tag{4.56}$$

Worksheets have been widely used in mobile e-commerce [39, 40] and are also suitable for edge computing applications such as internal due date assignment due to their low computational cost and cross-platform nature. In the cloud and edge computing system for internal due date assignment, the extracted RF is embedded into a worksheet to support internal due date assignment on smartphones, because decision rules can be easily implemented using cells with formulas (see Fig. 4.12). A set of cells with formulas constitutes a decision tree. Then, the RF is formed by grouping the set of cells in the worksheet. These worksheets are generated and saved in another cloud space every time the DNN is retrained.

When a customer places a potential order, the sales representative (i.e., the external edge node) downloads the edge computing application (i.e., the corresponding worksheet) for edge computing. The sales representative then applies the RF embedded in the edge computing application using a smart phone to assign an internal due date for the potential order. The sales representative can then negotiate the internal due date with the customer.

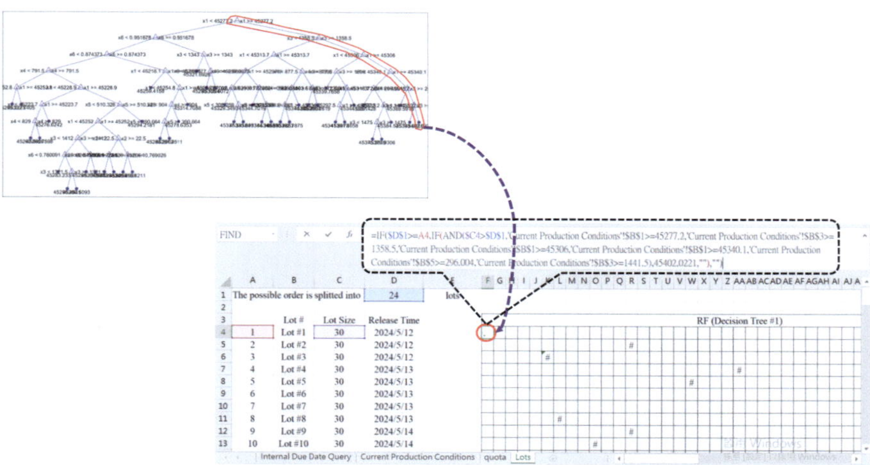

Fig. 4.12 Excel worksheets supporting edge computing applications [17]

4.5 Output Projection

4.5.1 Output Projection Using Traditional Methods

Inaccurate output prediction may lead to delivery delays, potential order loss, or unnecessary inventory accumulation, which is a critical but challenging task for manufacturing systems. Therefore, various ML or DL methods have been applied to output prediction.

For example, Chen and Wang [41] proposed a hybrid approach to estimate the monthly output of each product type in a semiconductor manufacturing plant. First, the fuzzy c-means (FCM) and FBPN methods were applied to estimate the completion time of each job in the plant. Then, the fuzzy output projection function (FOPF) was designed to project these outputs into each future time period. In order to evaluate the advantages and disadvantages of the hybrid approach, a production simulation was conducted.

To improve the effectiveness of monthly output projection, Chen [1] proposed a fuzzy neural method. First, the look-ahead SOM-FBPN method was used to estimate the completion time of each job in the factory. Then, the fuzzy monthly production function (FMOF) was designed to estimate the output for each future month. Take Fig. 4.13 as an example, in which the size and estimated completion time of order u are denoted by S_u and $\tilde{O}_u$, respectively. Then, the contributions of this order to the outputs of various months are

$$\text{Month } t{-}1\text{: } \left(\frac{1}{2}\left(\frac{t - O_{u1}}{O_{u2} - O_{u1}}\right) \cdot (t - O_{u1})\right) \cdot \frac{2S_u}{O_{u3} - O_{u1}}.$$

$$\begin{aligned}\text{Month } t\text{: } & \left(\frac{1}{2}(O_{u3} - O_{u1}) - \frac{1}{2}\left(\frac{t - O_{u1}}{O_{u2} - O_{u1}}\right) \cdot (t - O_{u1})\right. \\ & \left. - \frac{1}{2}\left(\frac{t + 1 - O_{u3}}{O_{u2} - O_{u3}}\right) \cdot (O_{u3} - t - 1)\right) \cdot \frac{2S_u}{O_{u3} - O_{u1}}.\end{aligned}$$

$$\text{Month } t + 1\text{: } \left(\frac{1}{2}\left(\frac{t + 1 - O_{u3}}{O_{u2} - O_{u3}}\right) \cdot (O_{u3} - t - 1)\right) \cdot \frac{2S_u}{O_{u3} - O_{u1}}.$$

The FMOF function is

$$O(t) = \sum_u S_u \cdot \frac{\int_t^{t+1} \mu_{\tilde{O}_u}(x)dx}{\int_0^{\infty} \mu_{\tilde{O}_u}(x)dx} \tag{4.57}$$

If $\tilde{O}_u$ is a triangular fuzzy number (TFN),

$$O(t) = \sum_u S_u \cdot \frac{f(\tilde{O}_u)}{\frac{O_{u3} - O_{u1}}{2}} \tag{4.58}$$

Fig. 4.13 Output projection based on a fuzzy completion time

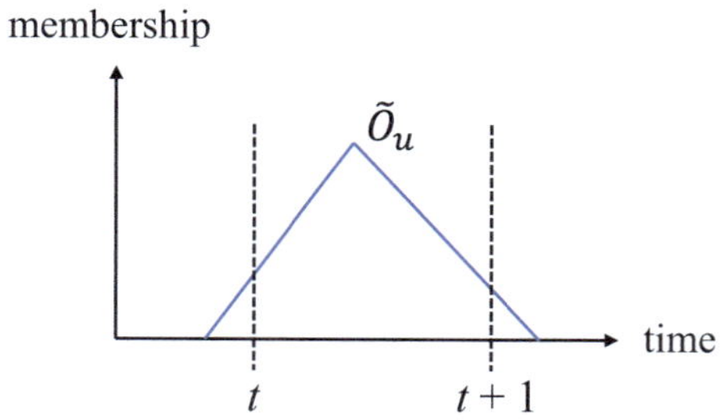

where

$$f(\tilde{O}_u)=\begin{cases} 0 & \text{if } O_{u3}\le t \\ \left(\frac{t+1-O_{u1}}{O_{u2}-O_{u1}}\right)\cdot\left(\frac{t+1-O_{u1}}{2}\right) & \text{if } t\le O_{u1}\le t+1\le O_{u2} \\ \left(\frac{t-O_{u1}}{O_{u2}-O_{u1}}+1\right)\cdot\frac{(O_{u2}-t)}{2}+\left(\frac{t+1-O_{u3}}{O_{u2}-O_{u3}}+1\right)\cdot\frac{(t+1-O_{u2})}{2} & \text{if } O_{u1}\le t\le O_{u2}\le t+1\le O_{u3} \\ \left(\frac{t-O_{u3}}{O_{u2}-O_{u3}}\right)\cdot\left(\frac{O_{u3}-t}{2}\right) & \text{if } O_{u2}\le t\le O_{u3}\le t+1 \\ \frac{O_{u2}-O_{u1}}{2}+\left(\frac{t+1-O_{u3}}{O_{u2}-O_{u3}}+1\right)\cdot\frac{(t+1-O_{u2})}{2} & \text{if } t\le O_{u1}\le O_{u2}\le t+1\le O_{u3} \\ \left(\frac{t-O_{u1}}{O_{u2}-O_{u1}}+1\right)\cdot\frac{(O_{u2}-t)}{2}+\frac{O_{u3}-O_{u2}}{2} & \text{if } O_{u1}\le t\le O_{u2}\le O_{u3}\le t+1 \\ \frac{O_{u3}-O_{u1}}{2} & \text{if } t\le O_{u1}\le O_{u2}\le O_{u3}\le t+1 \\ \left(\frac{t-O_{u1}}{O_{u2}-O_{u1}}+\frac{t+1-O_{u1}}{O_{u2}-O_{u1}}\right)\cdot\frac{(t+1-t)}{2} & \text{if } O_{u1}\le t\le t+1\le O_{u2} \\ \left(\frac{t-O_{u3}}{O_{u2}-O_{u3}}+\frac{t+1-O_{u3}}{O_{u2}-O_{u3}}\right)\cdot\frac{(t+1-t)}{2} & \text{if } O_{u2}\le t\le t+1\le O_{u3} \end{cases} \quad (4.59)$$

The job completion time estimated by the look-ahead SOM-FBPN method was an asymmetric TFN biased toward the side that contained the actual value. In theory, such a property made the proposed FMOF function superior to a crisp or probability function.

Example 4.6 The fuzzy completion times of ten orders have been estimated and are listed in Table 4.12. Based on these estimated completion times, the monthly output for each month in the future is to be projected. For example, the contribution of order #2 to the outputs of various months can be estimated as

$$\begin{aligned} &\text{Month } 6: 0\ (\text{pieces}) \\ &\text{Month } 7: 3\ (\text{pieces}) \\ &\text{Month } 8: 9\ (\text{pieces}) \\ &\text{Month } 9: 10\ (\text{pieces}) \\ &\text{Month } 10: 5\ (\text{pieces}) \\ &\text{Month } 11: 1\ (\text{pieces}) \end{aligned}$$

The output projection results are summarized in Table 4.13.

Leachman et al. [42] proposed the SLIM method to speed up the progress of certain wafer lots to achieve the monthly output target for Samsung Electronics Co.,

Table 4.12 Estimated fuzzy completion times of ten jobs

u	$\tilde{O}_u$ (month)	S_u (pieces)
1	(6.97, 9.08, 11.67)	28
2	(9.55, 10.86, 12.38)	16
3	(12.39, 13.38, 16.76)	27
4	(12.55, 16.88, 20.37)	29
5	(13.47, 16.58, 18.99)	21
6	(16.3, 17.77, 20.72)	14
7	(18.86, 22.78, 26.7)	14
8	(20.11, 24.27, 25.42)	15
9	(22.52, 23.92, 28.42)	22
10	(25.06, 25.88, 29.35)	19

Ltd. (SEC), one of the world's largest foundries. The cycle times of these wafer lots were thus shortened; however, the cycle times of other wafer lots were inevitably extended. The premise of the SLIM method is the accuracy of the output projection results. Otherwise, the inaccuracy of the output projection results will lead to an inappropriate job scheduling plan.

4.5.2 *Output Projection Using Industry 4.0 and Explainable Artificial Intelligence Methods*

Chen et al. [43] proposed an Industry 4.0 and explainable artificial intelligence (XAI) method to enhance the effectiveness and interpretability of output projection in wafer fabrication. In the Industry 4.0 and XAI approach, a DNN was first constructed to estimate the completion time of each job. Then, LIME was used to help interpret the trained DNN by approximating it with an RF. After evaluating the importance of each job attribute using Shapley value (SHAP) analysis, only the relevant decision trees in the RF were used to estimate the completion time of each job.

The Industry 4.0 and XAI approach consists of the following steps:

Step 1. Apply Industry 4.0 technologies (such as cloud computing, 5G, Internet of machines (IoM), and RFID) to collect data required for output projection.
Step 2. Build and train a DNN to estimate the completion time of each job.
Step 3. Use an RF to approximate the trained DNN (see Fig. 4.14): Traditionally, in LIME, linear regression (LR) equations are fitted to a local solution space using synthetic data, and an RF can be used to enhance its effectiveness.
Step 4. Use SHAP analysis to evaluate the importance of each job attribute.
Step 5. Apply only relevant decision trees in the RF to estimate the completion time of each job.
Step 6. Project monthly output.

Table 4.13 Output projection results

t	6	7	8	9	10	11	12	13	14	15	16	17	18	19	20	21	22	23	24	25	26	27	28
$O(t)$ (pieces)	0	3	9	11	13	8	3	13	14	15	16	16	11	6	3	4	7	13	12	11	10	7	2

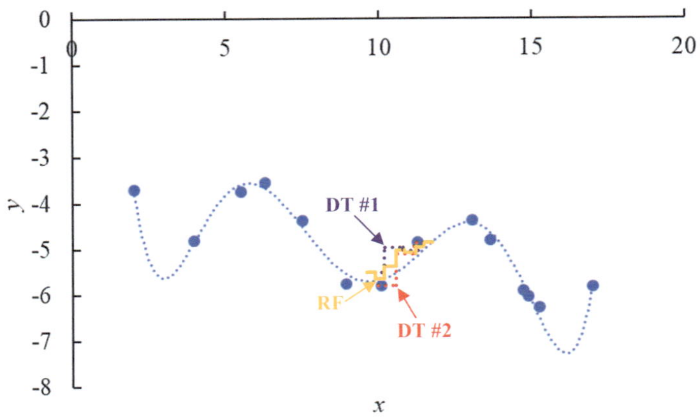

Fig. 4.14 Using an RF to approximate the trained DNN

Step 7. Evaluate the effectiveness of the Industry 4.0 and XAI approach.

The procedure is shown in Fig. 4.15.

4.5.3 Appling Industry 4.0 Technologies to Collect the Required Data

Some products (such as wafers) have very long manufacturing processes, and collecting real-time production condition information seems unrealistic. However, if any machine downstream in the manufacturing process fails, it may seriously affect the completion time of the job, thereby damaging the reliability of the output projection results. Therefore, it is necessary to apply Industry 4.0 technologies to collect or monitor the real-time status of these machines.

Wafer fabrication equipment, such as chemical delivery systems, cleaning equipment, chemical vapor deposition (CVD) and physical vapor deposition (PVD) equipment, and diffusion furnaces, all have embedded sensors and minicomputers and can therefore be considered cyber-physical systems (CPSs) [44]. SVID data collected by such sensors can be fed to the DNN for output projection and other purposes [45].

In addition, the real-time information about the status of a job (or logistics) can also be monitored. RFID tags attached to wafer carriers can also be used. Forked light barriers attached to a carrier, controlled by an ATMega32 microcontroller, detected the presence of a wafer that has completed fabrication and been ejected by the equipment [46]. In this way, the start and completion times of each job on a specific equipment can be known immediately and accurately. These data are also fed into the DNN for output projection.

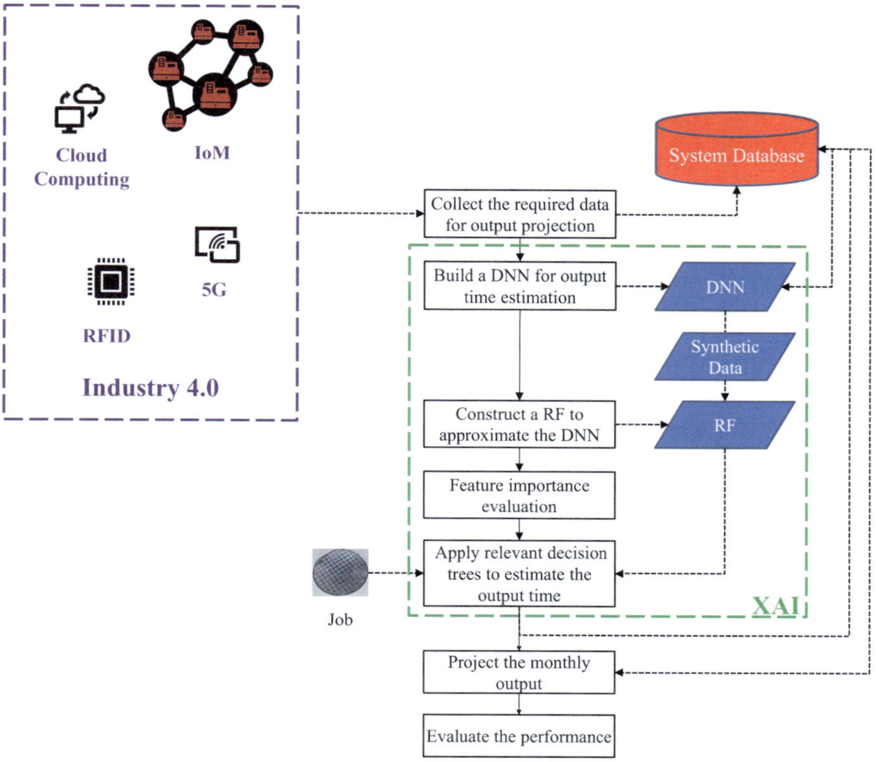

Fig. 4.15 Procedure of the Industry 4.0 and XAI approach

Detecting possible anomalies in wafer fabrication equipment based on data collected by embedded sensors can reduce quality issues by 35%, increase production capacity by 10%, and speed up yield learning by 25%, ultimately contributing to more monthly output [47]. Such collected data can also be used as inputs to the DNN for job completion estimation (or output projection).

4.5.4 Building and Training a DNN to Estimate Job Output Times

In the Industry 4.0 and XAI approach, at first, a DNN is built to estimate the completion time of a job based on job, machine and factory information collected by Industry 4.0 applications. The DNN is a feedforward neural network (FNN) with V hidden layers, aiming to minimize the mean squared error (MSE) in estimating the completion times of jobs in the training data:

$$\mathrm{MSE} = \frac{\sum_{j=1}^{n}(o_j - \hat{o}_j)^2}{n} \tag{4.60}$$

where

$$\hat{o}_j = \mathrm{DNN}(\{x_{jm}\}) \tag{4.61}$$

o_j and $\hat{o}$ denote the actual completion time and completion time forecast of job j; $j = 1 \sim n$. The most common algorithm for training a DNN is the LM algorithm [48].

Subsequently, following the concept of LIME, an RF is constructed to approximate the trained DNN (see Fig. 4.16), so as to find simpler decision rules suitable for various local regions of the sample space and avoid the overfitting problems that may occur in these regions [49].

There are L decision trees in RF. Each decision tree considers only randomly selected job attributes (i.e., feature bagging) and fits only randomly selected training example (i.e., bootstrapping) [49]. Each fitted decision tree has several decision rules in the following format:

$$\text{“If } x_{j1} \wedge_{k1}(l)\xi_{k1}(l) \text{ and} \ldots \text{and } x_{jM} \wedge_{kM}(l)\xi_{kM}(l) \text{ Then } \hat{o}_j(l) = c_k(l)\text{”} \tag{4.62}$$

where $\xi_{km}(l)$ indicates the threshold of x_{jm} in decision rule k of decision tree l; $k = 1 \sim K$; $l = 1 \sim L$. $\wedge_{km}(l) \in \{=, <, \geq, \mathit{within}\}$ is the logical operator; $c_k(l)$ is the outcome of this decision rule that is used to approximate the DNN output; and $c_k(l)$ is usually determined by averaging the DNN outputs of all jobs to which the decision rule applies (i.e., soft voting) [50, 51]. However, for output projection, the future values of many inputs (i.e., production conditions) required for cycle time prediction need to be predicted from their historical values:

$$\text{“If } x_{jm}(t-g) \wedge_{m1}(l)\xi_{m1}(l) \text{ and} \ldots \text{and } x_{jm}(t-1) \wedge_{mg}(l)\xi_{mg}(l) \text{ then } \hat{x}_{jm}(t) = \alpha_m(l)\text{”} \tag{4.63}$$

Then, Eq. (4.62) becomes

$$\text{“If } \hat{x}_{j1} \wedge_{k1}(l)\xi_{k1}(l) \text{ and} \ldots \text{and } \hat{x}_{jM} \wedge_{kM}(l)\xi_{kM}(l) \text{ then } \hat{o}_j(l) = c_k(l)\text{”} \tag{4.64}$$

4.5.5 Feature Importance Evaluation Using SHAP Analysis

The SHAP value of job attribute m is the weighted average of the marginal contributions in the context of all coalitions that exclude the job attribute [52]:

$$\varphi_m(f) = \sum_{\{\mathbf{S}\subseteq\mathbf{F}\}\{m\}} \frac{|\mathbf{S}|!(|\mathbf{F}| - |\mathbf{S}| - 1)!}{|\mathbf{F}|!} \circ \left[f\left(x_{\mathbf{S}\cup\{m\}}\right) - f(x_{\mathbf{S}})\right] \tag{4.65}$$

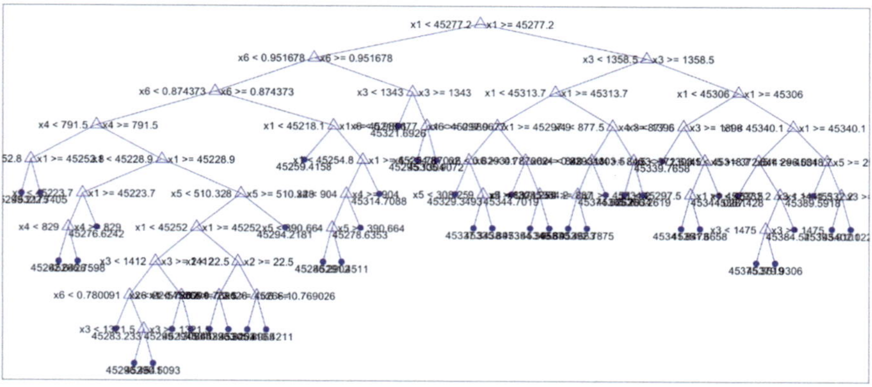

(Decision tree #1)

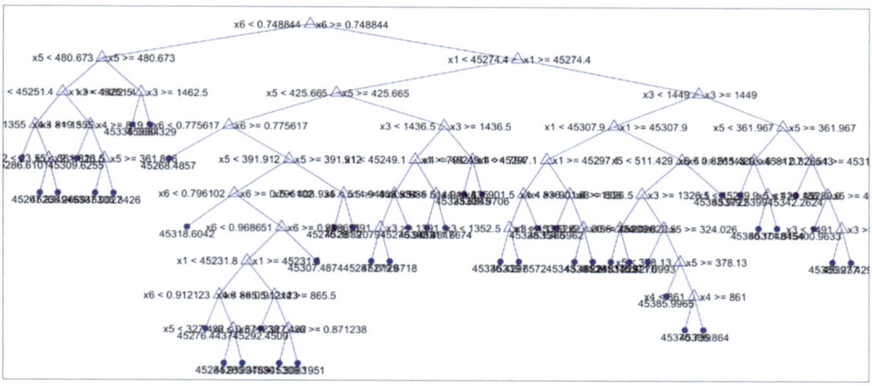

(Decision tree #2)

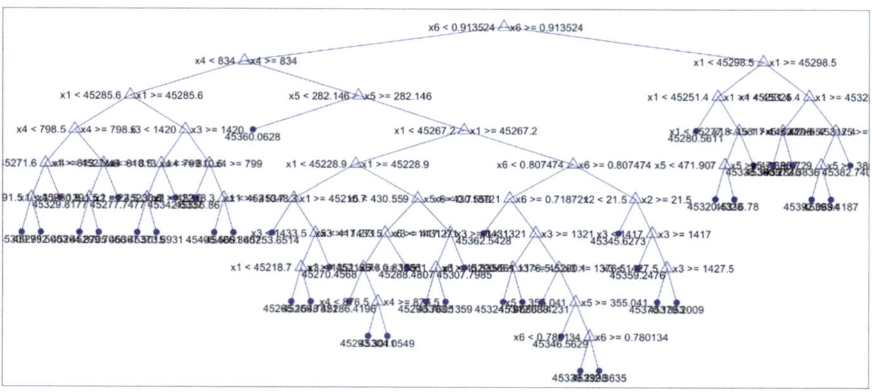

(Decision tree #*L*)

Fig. 4.16 Decision trees associated with the RF [17]

where $\varphi_m(f)$ is the weighted average Shapley value of attribute m for job j. $\mathbf{F}$ is the set of all job attributes, while $\mathbf{S}$ is its subset (or coalition). $f\left(x_{\mathbf{S}\cup\{m\}}\right)$ is the completion time estimation performance considering job attribute m, while $f(x_{\mathbf{S}})$ is the completion time estimation performance without considering job attribute m.

In short, SHAP analysis fixes the value of a job attribute and randomizes the values of other job attributes. The importance of a job attribute is directly proportional to the completion time estimation performance.

There are other ways to evaluate the importance of each job attribute [53], such as

- correlation analysis;
- partial derivative [54];
- odd ratio;
- out-of-bag (OOB) predictor importance;
- recursive feature elimination (RFE);
- learning important features (LIFT) [55, 56];
- deep learning important features (DeepLIFT) [57];
- deep SHAP [58];
- causal explanation (CXPlain) [59] as shown in Fig. 4.17.

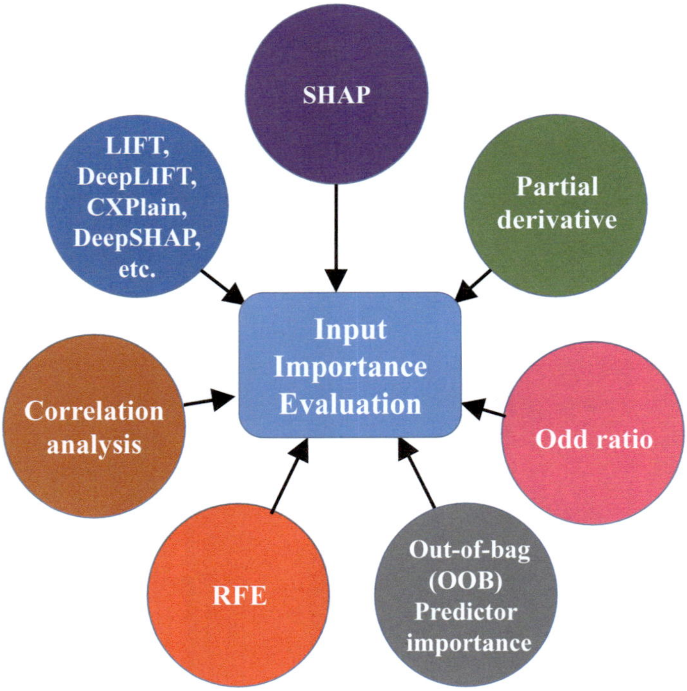

Fig. 4.17 Methods for evaluating the importance of each job attribute

4.5.6 Applying Relevant Decision Trees to Estimate the Job Completion Time

Accurate job completion time prediction is a premise of reliable output projection. Therefore, for job completion time estimation, job attributes with higher absolute SHAP values are more important [60–63], since the estimated completion times are as close to actual values as possible. The more important attributes for estimating the completion time of job j are put into set Ω_j:

$$\Omega_j = arg \max_m \left|\varphi_{jm}(f)\right| \tag{4.66}$$

On the other hand, if the number of nodes for the attribute (τ_{lm}) is more than the average, job attribute m is emphasized by the decision tree:

$$\tau_{lm} \geq \frac{1}{M} \sum_{\rho} \tau_{l\rho} \tag{4.67}$$

where τ_{lm} is the number of nodes for job attribute m in decision tree l.

Therefore, decision tree l can be applied to approximate the DNN output if all job attributes in Ω_j satisfy Constraint (4.62). The estimation results by eligible decision trees are then averaged:

$$\hat{o}_j = \sum_{\substack{m \in \Omega_j \\ \tau_{lm} \geq \frac{1}{M} \sum_{\rho} \tau_{l\rho}}} \overline{\hat{o}_j}(l) \tag{4.68}$$

4.5.7 Projecting the Monthly Output

The number of jobs expected to be outputted in a specific month determines the monthly output for the month:

$$O_t = \sum_{j=1}^{n} |\hat{o}_j : t \leq \hat{o}_j \langle t + 1| \tag{4.69}$$

which is the crisp output projection function (CPOF).

The output projection performance of the Industry 4.0 and XAI approach can be measured in terms of mean absolute error (MAE), mean absolute percentage error (MAPE), RMSE, and the coefficient of determination (R^2):

$$\mathrm{MAE} = \frac{\sum_{t=1}^{T} \left| O_t - \hat{O}_t \right|}{n} \tag{4.70}$$

$$\mathrm{MAPE} = \frac{1}{T} \frac{\sum_{t=1}^{T} \left| O_t - \hat{O}_t \right|}{O_t} \tag{4.71}$$

$$\mathrm{RMSE} = \sqrt{\frac{\sum_{t=1}^{T} (O_t - \hat{o}_t)^2}{T}} \tag{4.72}$$

$$R^2 = 1 - \frac{\sum_{t=1}^{T} (O_t - \hat{O}_t)^2}{\sum_{t=1}^{T} (O_t - \overline{O})^2} \tag{4.73}$$

where O_t is the actual monthly output within month t.

References

1. T. Chen, A fuzzy-neural and multiple-bucket approach for estimating lot cycle time in a wafer fab with dynamic product mix. Comput. Ind. Eng. **55**, 423–438 (2008)
2. T. Chen, A. Jeang, Y.-C. Wang, A hybrid neural network and selective allowance approach for internal due date assignment in a wafer fabrication plant. Int. J. Adv. Manuf. Technol. **36**, 570–581 (2008)
3. T. Chen, H.-C. Wu, A fuzzy-neural approach for output projection in a semiconductor fabrication factory. J. Chin. Inst. Eng. **32**(2), 287–293 (2009)
4. T. Chen, Y.-C. Wang, A nonlinear scheduling rule incorporating fuzzy-neural remaining cycle time estimator for scheduling a semiconductor manufacturing factory. Int. J. Adv. Manuf. Technol. **45**, 110–121 (2009)
5. Z.H. Che, T.A. Chiang, Designing a collaborative supply-chain plan using the analytic hierarchy process and genetic algorithm with cycle-time estimation. Int. J. Prod. Res. **50**(16), 426–4443 (2012)
6. S. Mosallaeipour, M.G. Nejad, S.M. Shavarani, R. Nazerian, Mobile robot scheduling for cycle time optimization in flow-shop cells, a case study. Prod. Eng. Res. Devel. **12**(1), 83–94 (2018)
7. Y. Zhao, X. Xu, H. Li, Y. Liu, Stochastic customer order scheduling with setup times to minimize expected cycle time. Int. J. Prod. Res. **56**(7), 2684–2706 (2018)
8. W. Shou, J. Wang, P. Wu, X. Wang, H.Y. Chong, A cross-sector review on the use of value stream mapping. Int. J. Prod. Res. **55**(13), 3906–3928 (2017)
9. X. Tan, L. Xing, Z. Cai, G. Wang, Analysis of production cycle-time distribution with a big-data approach. J. Intell. Manuf. **31**, 1889–1897 (2020)
10. V. Vinod, R. Sridharan, Simulation modeling and analysis of due-date assignment methods and scheduling decision rules in a dynamic job shop production system. Int. J. Prod. Econ. **129**(1), 127–146 (2011)
11. B.C. Choi, M.J. Park, Just-in-time scheduling with generalized due dates and identical due date intervals. Asia-Pacific J. Oper. Res. **35**(06), 1850046 (2018)
12. M. Atsmony, G. Mosheiov, Common due-date assignment problems with fixed-plus-linear earliness and tardiness costs. Comput. Ind. Eng. **188**, 109915 (2024)
13. R.J. Patil, Due date management to improve customer satisfaction and profitability. Int J Log Res Appl **13**(4), 273–289 (2010)
14. C.C. Wu, D. Bai, X. Zhang, S.R. Cheng, J.C. Lin, Z.L. Wu, W.C. Lin, A robust customer order scheduling problem along with scenario-dependent component processing times and due dates. J. Manuf. Syst. **58**, 291–305 (2021)

15. Y.W. Sang, J.Q. Wang, M. Sterna, J. Błażewicz, Single machine scheduling with due date assignment to minimize the total weighted lead time penalty and late work. Omega **121**, 102923 (2023)
16. S.C. Leung, Y. Wu, K.K. Lai, Multi-site aggregate production planning with multiple objectives: a goal programming approach. Prod. Plan. Control **14**(5), 425–436 (2003)
17. T.C.T. Chen, C.W. Lin, Feasibility of a clouds and edge computing scheme for internal due date assignment in advanced manufacturing systems. Int. J. Adv. Manuf. Technol. **136**(5), 2893–2908 (2025)
18. A. Romanov, A. Filippov, N. Yarushkina, Extraction and forecasting time series of production processes, in *International Conference on Information Technologies* (2019), pp. 173–184
19. T. Chen, The FLR-PCFI-RBF approach for accurate and precise WIP level forecasting. Int. J. Adv. Manuf. Technol. **63**, 1217–1226 (2012)
20. Y.-C. Wang, T.-C.T. Chen, M.-C. Chiu, An explainable deep-learning approach for job cycle time prediction. Decis. Analytics **6**, 100153 (2023)
21. V. Gordon, J.M. Proth, C. Chu, A survey of the state-of-the-art of common due date assignment and scheduling research. Eur. J. Oper. Res. **139**(1), 1–25 (2002)
22. S. Karsoliya, Approximating number of hidden layer neurons in multiple hidden layer BPNN architecture. Int. J. Eng. Trends Technol. **3**(6), 714–717 (2012)
23. B. Chen, J. Wan, A. Celesti, D. Li, H. Abbas, Q. Zhang, Edge computing in IoT-based manufacturing. IEEE Commun. Mag. **56**(9), 103–109 (2018)
24. Y. Feng, Z. Hong, Z. Li, H. Zheng, J. Tan, Integrated intelligent green scheduling of sustainable flexible workshop with edge computing considering uncertain machine state. J. Clean. Prod. **246**, 119070 (2020)
25. Z. Zhao, P. Lin, L. Shen, M. Zhang, G.Q. Huang, IoT edge computing-enabled collaborative tracking system for manufacturing resources in industrial park. Adv. Eng. Inform. **43**, 101044 (2020)
26. J. Zhang, C. Deng, P. Zheng, X. Xu, Z. Ma, Development of an edge computing-based cyber-physical machine tool. Robot. Comput.-Integr. Manuf. **67**, 102042 (2021)
27. Y. Wang, M. Liu, P. Zheng, H. Yang, J. Zou, A smart surface inspection system using faster R-CNN in cloud-edge computing environment. Adv. Eng. Inform. **43**, 101037 (2020)
28. Z. Zhu, G. Han, G. Jia, L. Shu, Modified densenet for automatic fabric defect detection with edge computing for minimizing latency. IEEE Internet Things J. **7**(10), 9623–9636 (2020)
29. Q. Qi, F. Tao, A smart manufacturing service system based on edge computing, fog computing, and cloud computing. IEEE Access **7**, 86769–86777 (2019)
30. R. Bryll, R. Gutierrez-Osuna, F. Quek, Attribute bagging: improving accuracy of classifier ensembles by using random feature subsets. Pattern Recogn. **36**(6), 1291–1302 (2003)
31. M.E.H. Daho, M.A. Chikh, Combining bootstrapping samples, random subspaces and random forests to build classifiers. J. Med. Imaging Health Inform. **5**(3), 539–544 (2015)
32. M.R. Zafar, N. Khan, Deterministic local interpretable model-agnostic explanations for stable explainability. Mach. Learn. Knowl. Extr. **3**(3), 525–541 (2021)
33. J.S. Chou, K.E. Chen, Optimizing investment portfolios with a sequential ensemble of decision tree-based models and the FBI algorithm for effective financial indicator analysis. Appl. Soft Comput. 111550 (2024)
34. V.G. Costa, S. Salcedo-Sanz, C.E. Pedreira, Efficient evolution of decision trees via fully matrix-based fitness evaluation. Appl. Soft Comput. **150**, 111045 (2024)
35. V. Rodriguez-Galiano, M. Sanchez-Castillo, M. Chica-Olmo, M.J.O.G.R. Chica-Rivas, Machine learning predictive models for mineral prospectivity: an evaluation of neural networks, random forest, regression trees and support vector machines. Ore Geol. Rev. **71**, 804–818 (2015)
36. S.S. Rathore, S. Kumar, A decision tree regression based approach for the number of software faults prediction. ACM SIGSOFT Softw. Eng. Notes **41**(1), 1–6 (2016)
37. H. Pham, S. Olafsson, On Cesaro averages for weighted trees in the random forest. J. Classif. **37**(1), 223–236 (2020)
38. T. Chen, Y.C. Wang, H.C. Wu, Analyzing the impact of vaccine availability on alternative supplier selection amid the COVID-19 pandemic: a cFGM-FTOPSIS-FWI approach. Healthcare **9**(1), 71 (2021)

39. P. Wu, F. Xiong, Q. Cao, Research on mobile e-commerce model oriented manufacturing service, in *International Conference on Computer, Communications and Information Technology* (2014), pp. 5–7
40. T. Chen, M.C. Chiu, Smart technologies for assisting the life quality of persons in a mobile environment: a review. J. Ambient. Intell. Humaniz. Comput. **9**, 319–327 (2018)
41. T. Chen, Y.-C. Wang, A hybrid intelligent approach for output projection in a semiconductor fabrication plant. Intell. Data Anal. **12**(1), 129–144 (2008)
42. R.C. Leachman, J. Kang, V. Lin, SLIM: short cycle time and low inventory in manufacturing at Samsung electronics. Interfaces **32**(1), 61–77 (2002)
43. T.C.T. Chen, Y.C. Wang, C.W. Lin, Enhancing the effectiveness of output projection in wafer fabrication using an industry 4.0 and XAI approach. Int. J. Adv. Manuf. Technol. **134**(1), 113–125 (2024)
44. PEER Group Inc., SEMI PV2 (2022). https://www.peergroup.com/expertise/resources/semistandards/pv2/
45. T.C.T. Chen, Industry 4.0 for semiconductor manufacturing, in *Production Planning and Control in Semiconductor Manufacturing: Big Data Analytics and Industry 4.0 Applications* (2022), pp. 21–40
46. D.H. Lee, C.H. Lee, S.H. Choi, K.J. Kim, A method for wafer assignment in semiconductor wafer fabrication considering both quality and productivity perspectives. J. Manuf. Syst. **52**, 23–31 (2019)
47. L.C. Wang, P.C. Chu, S.Y. Lin, Impact of capacity fluctuation on throughput performance for semiconductor wafer fabrication. Robot. Comput.-Integr. Manuf. **55**, 208–216 (2019)
48. B.M. Wilamowski, H. Yu, Improved computation for Levenberg–Marquardt training. IEEE Trans. Neural Networks **21**(6), 930–937 (2010)
49. H. Prabhu, A. Sane, R. Dhadwal, N.R. Parlikkad, J.K. Valadi, Interpretation of drop size predictions from a random forest model using local interpretable model-agnostic explanations (LIME) in a rotating disc contactor. Ind. Eng. Chem. Res. **62**(45), 19019–19034 (2023)
50. T.C.T. Chen, H.C. Wu, M.C. Chiu, A deep neural network with modified random forest incremental interpretation approach for diagnosing diabetes in smart healthcare. Appl. Soft Comput. **152**, 111183 (2024)
51. S. Kumari, D. Kumar, M. Mittal, An ensemble approach for classification and prediction of diabetes mellitus using soft voting classifier. Int. J. Cogn. Comput. Eng. **2**, 40–46 (2021)
52. F. Gabbay, S. Bar-Lev, O. Montano, N. Hadad, A lime-based explainable machine learning model for predicting the severity level of covid-19 diagnosed patients. Appl. Sci. **11**(21), 10417 (2021)
53. T.C.T. Chen, XAI applications in job sequencing and scheduling, in *Explainable and Customizable Job Sequencing and Scheduling: Advancing Production Control and Management with XAI* (2025), pp. 37–71
54. M. Green, U. Ekelund, L. Edenbrandt, J. Björk, J.L. Forberg, M. Ohlsson, Exploring new possibilities for case-based explanation of artificial neural network ensembles. Neural Netw. **22**(1), 75–81 (2009)
55. H. Hamilton, Cumulative gains and lift charts (2009). https://www2.cs.uregina.ca/~dbd/cs831/notes/lift_chart/lift_chart.html
56. B. Keating, Lift charts (2024). https://www3.nd.edu/~busiforc/handouts/DataMining/Lift%20Charts.html
57. A. Shrikumar, P. Greenside, A. Kundaje, Learning important features through propagating activation differences, in *International Conference on Machine Learning* (2017), pp. 3145–3153
58. H. Chen, S.M. Lundberg, S.I. Lee, Explaining a series of models by propagating Shapley values. Nat. Commun. **13**(1), 4512 (2022)
59. P. Schwab, W. Karlen, Cxplain: causal explanations for model interpretation under uncertainty, in *Advances in Neural Information Processing Systems* (2019), pp. 1–11
60. A. Dikshit, B. Pradhan, Interpretable and explainable AI (XAI) model for spatial drought prediction. Sci. Total. Environ. **801**, 149797 (2021)

61. H.A. Al-Najjar, B. Pradhan, G. Beydoun, R. Sarkar, H.J. Park, A. Alamri, A novel method using explainable artificial intelligence (XAI)-based Shapley additive explanations for spatial landslide prediction using time-series SAR dataset. Gondwana Res. **123**, 107–124 (2023)
62. M.C. Chiu, T.C.T. Chen, Y.C. Wang, Flexible and objective diagnosis of type II diabetes by using a fuzzy deep learning ensemble approach. Complex Intell. Syst. **11**(6), 1–15 (2025)
63. T.C.T. Chen, C.W. Lin, A hybrid XAI-FCI approach for job cycle time range estimation: wafer fabrication as an example. Oper. Res. Int. J. **25**(2), 1–29 (2025)

Chapter 5
Cycle Time Reduction

Abstract For many manufacturing systems, cycle time reduction is obviously the key to improving competitiveness. Therefore, cycle time reduction is regarded as the most critical and urgent task in production control. This chapter first summarizes the benefits of cycle time reduction on manufacturing systems from existing references and practices, such as reducing work-in-process levels, improving productivity, reducing unit costs, responding to market demand changes, and improving yield. This chapter also points out that cycle time reduction does not necessarily mean speeding up the execution of all operations. In addition, the cycle time reduction target of a product can be decomposed into the targets of its individual components based on the results of the bill of materials or material requirement planning. Subsequently, some methods for reducing cycle time are introduced, including more efficient job scheduling, yield improvement, controlling production conditions, 3D printing, leveraging cloud manufacturing capabilities, pull manufacturing, value stream mapping, and lean six sigma. Obviously, cycle time can be reduced in all aspects of system planning, control, and management. Finally, this chapter also provides numerical examples to illustrate the application of these cycle time reduction methods and discusses their potential effects.

Keywords Cycle time reduction · Cycle time · Competitiveness · Responsiveness · Value stream mapping · 3D printing

5.1 Cycle Time Reduction

The most significant benefit of cycle time reduction is that the inventory (i.e., work-in-progress, WIP) level in the factory decreases because jobs do not stay in the factory as long [1]. Since inventory represents an opportunity cost, a reduction in cycle time also reflects a reduction in the unit cost. In addition, through cycle time reduction, the factory can better respond to changes in market demand [2]. A particular advantage of cycle time reduction is reflected in wafer fabrication plants (wafer fabs), in which the risk of wafer contamination increases as cycle time lengthens [2]. Therefore,

T.-C. T. Chen, *Advances in Cycle Time Management*,
https://doi.org/10.1007/978-3-032-06911-5_5

cycle time reduction helps to improve the yield of wafers. The benefits of cycle time reduction are summarized in Fig. 5.1.

In the literature, Taifa and Vhora [3] linked cycle time reduction to increased productivity. In theory, productivity can be measured as output divided by the inputs (including time and other resources). Therefore, increasing output or reducing inputs can increase productivity. If productivity is high, more output and profits can be generated from the same amount of inputs. More profits expand the room for flexible pricing, making it easier to expand the market share, thereby improving the competitiveness of the manufacturing system (see Fig. 5.2). Cycle time reduction is also linked to quick response. Shortened cycle times also result in faster delivery, allowing companies to respond quickly.

Chen [4] also emphasized the importance of cycle time reduction for dynamic random access memory (DRAM) manufacturers to respond to market demand fluctuations. However, cycle time reduction in the semiconductor industry is a challenging task. As the focal company of the semiconductor supply chain, numerous logistical factors affect the ability of a wafer fabrication plant (wafer fab) to control cycle time.

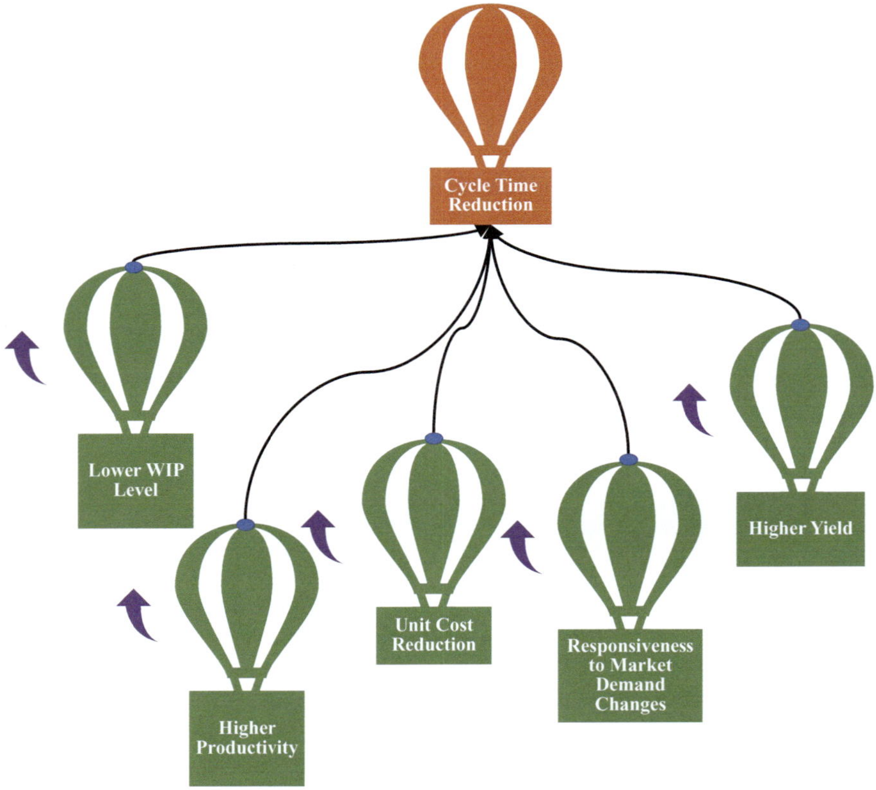

Fig. 5.1 Benefits of cycle time reduction

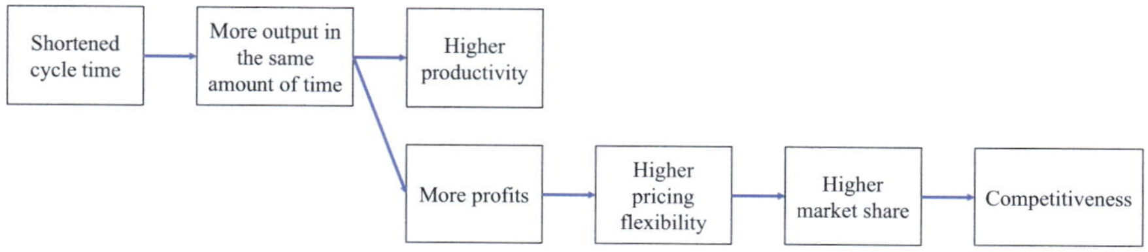

Fig. 5.2 Linking cycle time reduction to productivity improvement and competitiveness

After procuring raw materials, the release strategy has a greater impact on cycle time than the scheduling method. Many DRAM manufacturers release jobs to their fabs in a uniform manner and sequence them according to a prioritized first-in, first-out strategy (i.e., jobs are sorted by priorities first, then by their arrival times) [5]. Despite this, there are still some studies that try to reduce cycle time by improving job scheduling [6].

5.2 Cycle Time Reduction Methods

Methods for cycle time reduction can be divided into several categories (see Fig. 5.3):

- More effective job scheduling.
- Yield improvement.
- Controlling production conditions.
- Three-dimensional (3D) printing: Attaran [1] highlighted the potential of 3D printing in shortening cycle times. Since 3D printing can be used to manufacture finished products directly from raw materials without having to go through the steps one by one, cycle times can be significantly shortened. In addition, 3D printing makes mass customization very straightforward. In contrast, traditional production methods require a lot of time to prepare different materials and molds and re-set machines, resulting in very long cycle times.
- Utilizing cloud manufacturing capacity.
- Pull manufacturing.
- Value stream mapping (VSM).

Some studies [3] advocate that all operations of a job should be quickened to shorten the cycle time of the job. However, such efforts can be selective. Namely, it is not necessary to quicken all operations of a job for cycle time reduction. As the famous project management tool, critical path method (CPM) emphasizes only the operations on the critical path will affect the completion time (and cycle time) of the job (see Fig. 5.4).

Similarly, the cycle time reduction target of a product can be broken down into the targets for each component of the product based on the bill of materials (BOM) of the product or material requirement planning (MRP) results. For example, in Fig. 5.5,

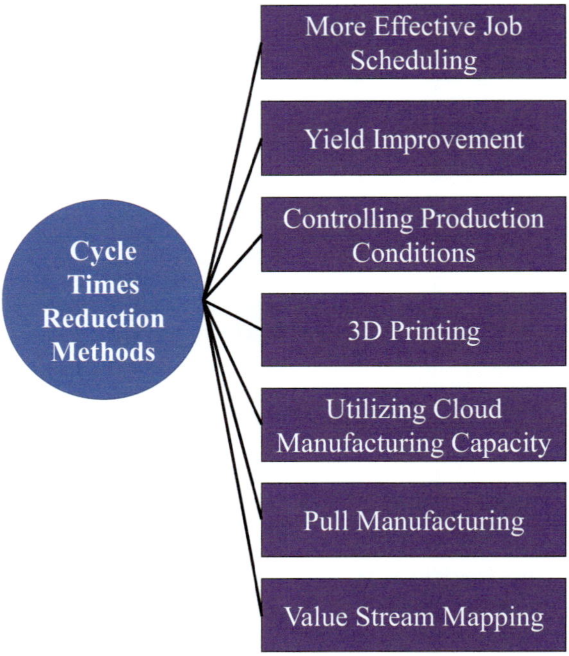

Fig. 5.3 Cycle time reduction methods

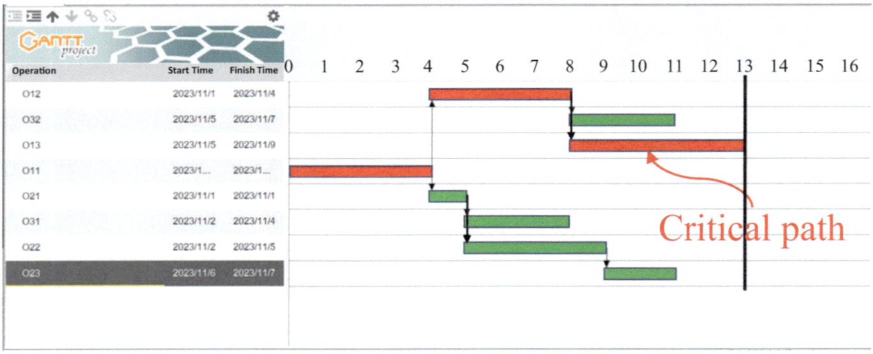

Fig. 5.4 Operations of a job on the critical path

three units of component A1 and four units of component A2 make up one unit of product A. Since the two components can be produced simultaneously, the cycle time for product A is max(90, 68) + 40 = 130 (min). To shorten the cycle of product A by 20 min, one way is to shorten the time to assemble all components from 40 min to 40 − 20 = 20 min, which is very challenging. Another way is to shorten the cycle time to complete all components from max(90, 68) min to max(90, 68) − 20

= 70 min, which is equivalent to shortening the time to produce each component A1 from 30 min to 70/3 = 23.3 min, which may be much easier.

Other cycle time reduction problems can be more challenging, as shown in the following example.

Example 5.1 The BOM of product B is shown in Fig. 5.6. The cycle time for making each component is provided in Table 5.1. As a result, the cycle time of product B can be derived as

$$C_B = \max((3 \cdot 25 + 20) \cdot 2, \max(2 \cdot 25, 4 \cdot 40) + 20) + 40 = 230 \text{ (min)}$$

To reduce the cycle time of product B by 10% to $230 \cdot (1 - 10\%) = 207$ min, the cycle times on the left and right sides of Fig. 5.6 need to be shortened by 230 − 207 = 23 min and 220 − 207 = 13 min, respectively. First, on the left side, the cycle

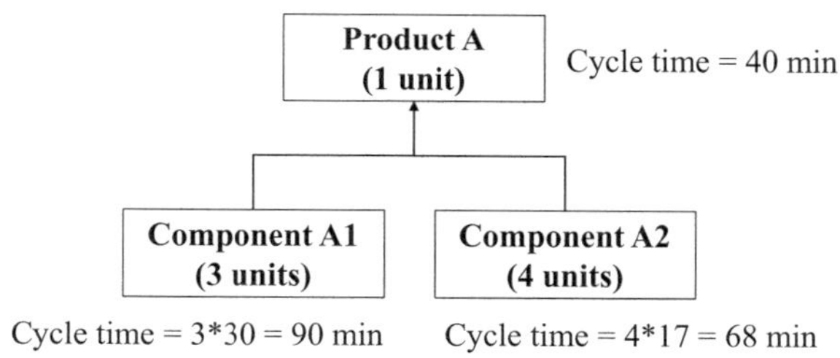

Fig. 5.5 Breaking down the cycle time reduction target of a product into the target for each component

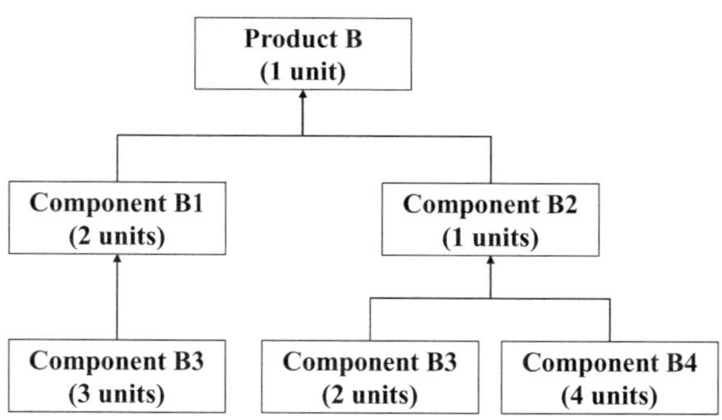

Fig. 5.6 BOM of product B

Table 5.1 Cycle time for making each component

j (component)	C_j (min.)
B	40
B1	20
B2	20
B3	25
B4	40

Table 5.2 Cycle time reduction plan

j (component)	C_j (min.)
B	40
B1	20
B2	20
B3	25 → 21.2
B4	40 → 36.8

time of B1 can be reduced by 23/2 = 11.5 min, which is up to 11.5/20 = 57.5%. Another way is to shorten the cycle time of B3 by 23/(2·3) = 3.8 min (with a change of 3.8/25 = 15%) to 25 − 3.8 = 21.2 (min). As a result, the cycle time on the right side remains 220 min. Therefore, the cycle time of B2 needs to be reduced to 20 − 13 = 7 (with a change of up to 7/20 = 35%). Another way is to shorten the cycle time of B4 to 40 − 13/4 = 36.8 (with a change of only 3.3/40 = 8%). Finally, the cycle time reduction plan is formulated and presented in Table 5.2.

5.3 More Effective Job Sequencing and Scheduling

5.3.1 Cycle Time Reduction Methods in Simple Manufacturing Systems

Job sequencing and scheduling is one of the most commonly applied methods for reducing cycle times. For some simple manufacturing systems, dispatching rules that minimize cycle times have been proved theoretically (see Table 5.3) [7].

Table 5.3 Some dispatching rules for reducing cycle times

Scheduling problem	Dispatching rule	Job priority evaluation formula
$1//\Sigma C_j$	Shortest processing time (SPT)	p_j
$1//\Sigma w_j C_j$	Weighted shortest processing time (WSPT)	p_j/w_j
$1/\text{prec}/\Sigma w_j C_j$	ρ-factor (largest ratio of weight and processing time)	$\max\limits_{1 \le l \le k} \frac{\sum_{j=1}^{l} w_j}{\sum_{j=1}^{l} p_j}$
$1/r_j/C_{\max}$	Earliest due date (EDD)	$d_j = K - r_j$; $K > \max\limits_j r_j$
$F2//C_{\max}$	Johnson's algorithm	p_{1j}; $p_{1j} < p_{2j}$ $1/p_{2j}$; $p_{1j} \ge p_{2j}$

5.3.2 Linear, Nonlinear, and Optimized Fluctuation Smooth Rules

Two fluctuation smoothing (FS) rules have been proposed to schedule jobs in a wafer fab to minimize the average cycle time and cycle time variation [8–10]. The first is the fluctuation smoothing rule for mean cycle time (FSMCT) [8]:

$$\mathrm{SK}_{jm} = j/\lambda - \mathrm{RCTE}_{jm} \tag{5.1}$$

where SK_{jm} is the slack of job j to be processed by machine m; λ is the release rate; RCTE_{jm} is the estimated remaining cycle time of job j since the operation at machine m. RCTE_{jm} can be derived by subtracting the current time (t) from the estimated completion time that is the release time (R_j) plus the cycle time forecast (CTE_j):

$$\mathrm{RCTE}_{jm} = R_j + \mathrm{CTE}_j - t \tag{5.2}$$

The other is the fluctuation smoothing rule for cycle time variation (FSVCT):

$$\mathrm{SK}_{jm} = R_j - \mathrm{RCTE}_{jm} \tag{5.3}$$

In both scheduling rules, the (remaining) cycle time of each job needs to be predicted, which shows the importance of cycle time prediction for cycle time reduction.

Example 5.2 The example from [10] is used to illustrate the application of the two FS rules, involving ten jobs to be sequenced (see Table 5.4), where the cycle time of each job is estimated using an artificial neural network (ANN); $t = 1204$ (minutes). The application results are summarized in Table 5.5. If the average cycle time is to be minimized, the jobs are processed in the following order: #89 → #108 → #118 → #115 → #131 → #153 → #158 → #141 → #132. On the contrary, if cycle time variation is to be minimized, the processing order is #89 → #158 → #150 → #115 → #153 → #108 → #132 → #131 → #141.

Chen [9] tailored the FSVCT rule to a specific wafer fab to further reduce the average cycle time. First, both terms in the FSVCT rule are normalized as

$$N(R_j) = \frac{R_j - \min_l R_l}{\max_l R_l - \min_l R_l} \tag{5.4}$$

$$N(\mathrm{RCTE}_{jm}) = \frac{\mathrm{RCTE}_{jm} - \min_l \mathrm{RCTE}_{lm}}{\max_l \mathrm{RCTE}_{lm} - \min_l \mathrm{RCTE}_{lm}} \tag{5.5}$$

Subsequently, to improve the responsiveness of the FSVCT rule, the division operator is adopted instead of the traditional subtraction operator:

Table 5.4 Data of ten jobs to be scheduled

j	R_j (min.)	CTE_j (min.)
118	935	1285
89	597	1491
158	921	1481
132	377	1368
131	788	1305
150	454	1470
141	612	1301
115	708	1448
108	770	1392
153	988	1406

$$\text{SK}_{jm} = \frac{N(R_j)}{N(\text{RCTE}_{jm})} \tag{5.6}$$

which becomes a nonlinear FSVCT rule. Nonlinear FSMCT is derived similarly as

$$\text{SK}_{jm} = \frac{N\left(\frac{j}{\lambda}\right)}{N(\text{RCTE}_{jm})} \tag{5.7}$$

Example 5.3 The two nonlinear FS rules are applied to the case in Example 5.2 ($\lambda = 1$), and the results are shown in Table 5.6. In this case, the sequencing results using the linear and nonlinear FS rules are not the same (see Fig. 5.7).

To optimize the performance of the nonlinear FSVCT rule for a factory, Chen [11] adds two coefficients, ξ and ζ, to the nonlinear FSVCT rule [10]:

$$\begin{aligned}\text{SK}_{jm}(\xi, \zeta) = {} & \left(\frac{\alpha\left(\text{RCTE}_{jm} - \min_l \text{RCTE}_{lm}\right)}{\beta_m}\right)^{-\xi} \\ & \cdot \left(R_j - \text{RCTE}_{jm} + (\text{RCTE}_{jm} - \min_l R_l)^{\zeta}\right)\end{aligned} \tag{5.8}$$

where $\alpha = \max_l R_l - \min_l R_l$; $\beta_m = \max_l \text{RCTE}_{lm} - \min_l \text{RCTE}_{lm}$; and ξ and ζ are positive real numbers satisfying the following constraints:

- $\xi = 0 \leftrightarrow \zeta = 0$.
- $\xi = 1 \leftrightarrow \zeta = 1$.

Theorem 5.1

- *Equation* (5.8) *becomes the traditional (linear) FSVCT when both* ξ *and* ζ *are equal to* 0.

Table 5.5 Application results of FSMCT and FSVCT

j	R_j (min.)	CTE$_j$ (min.)	RCTE$_j$ (min.)	SK$_{jm}$ (FSMCT)	Order (FSMCT)	SK$_{jm}$ (FSVCT)	Order (FSVCT)
118	935	1285	1016	719	3	− 81	10
89	597	1491	884	425	1	− 287	1
158	921	1481	1198	1126	7	− 277	2
132	377	1368	541	1400	9	− 164	7
131	788	1305	889	1037	5	− 101	8
150	454	1470	720	1486	10	− 266	3
141	612	1301	709	1365	8	− 97	9
115	708	1448	952	739	4	− 244	4
108	770	1392	958	630	2	− 188	6
153	988	1406	1190	1060	6	− 202	5

Table 5.6 Application results using nonlinear FSMCT and nonlinear FSVCT

j	R_j (min.)	CTE_j (min.)	RCTE_j (min.)	SK_{jm} (NFSMCT)	Order (NFSMCT)	SK_{jm} (NFSVCT)	Order (NFSVCT)
118	935	1285	1016	0.581	3	1.263	8
89	597	1491	884	0.000	1	0.690	3
158	921	1481	1198	1.000	6	0.890	5
132	377	1368	541	inf	10	0.000	1
131	788	1305	889	1.149	7	1.270	9
150	454	1470	720	3.244	9	0.462	2
141	612	1301	709	2.946	8	1.504	10
115	708	1448	952	0.602	4	0.866	4
108	770	1392	958	0.434	2	1.013	7
153	988	1406	1190	0.939	5	1.012	6

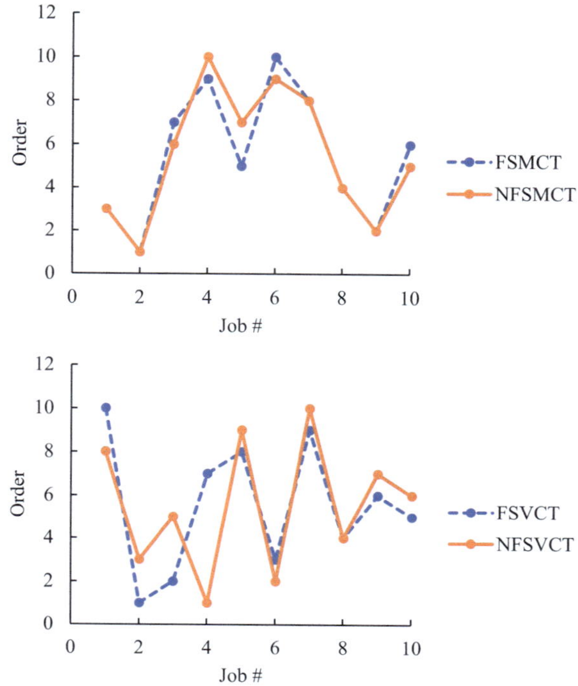

Fig. 5.7 Comparison of the sequencing results using the linear and nonlinear FS rules

- *Equation* (5.9) *is equivalent to the nonlinear FSVCT when both* ξ *and* ζ *are equal to* 1.

There are many possible models to choose the values of ξ and ζ in Eq. (5.8). For example,

- Linear model: $\xi = \zeta$.
- Nonlinear model: $\xi = \zeta^k, k \geq 0$.
- Logarithmic model: $\xi = \ln(1 + \zeta) / \ln 2$.

The values of ξ and ζ can be chosen to minimize cycle time variation, for which production simulation needs to be conducted to validate the optimality of the scheduling rule. Such settings are tailored to the factory.

The performance of the nonlinear FSMCT rule can be optimized for a factory similarly [10]:

$$\mathrm{SK}_{jm}(\xi, \zeta) = \beta_m(\frac{j}{\lambda} - \mathrm{RCTE}_{jm} + \gamma(\min_l \mathrm{RCTE}_{lm} - \frac{j}{\lambda}))^{-\eta} \tag{5.9}$$

where

$$\beta_m = \begin{cases} 1 & \text{if} \quad \gamma = 0 \\ -(j-1)\left(\max\limits_{l} \text{RCTE}_{lm} - \min\limits_{l} \text{RCTE}_{lm}\right)/(n-1) & \text{otherwise} \end{cases} \tag{5.10}$$

In addition, γ and η are real numbers satisfying the following constraints:

- $\gamma = 0 \leftrightarrow \eta = -1$.
- $\gamma = 1 \leftrightarrow \eta = 1$.

There are many possible models to choose the values of γ and η in Eq. (5.9). For example,

- Linear model: $\eta = 2\gamma - 1$.
- Nonlinear model: $\eta = 2\gamma^k - 1, k \geq 0$.
- Logarithmic model: $\eta = \ln(0.5 + 1.5\gamma)/\ln 2$.

5.3.3 *Other Scheduling Methods*

Mosallaeipour et al. [12] scheduled the operations of a robot for three machines in a cell production system to minimize the average cycle time. Each machine was equipped with a one-capacitated buffer. Therefore, it was not possible to accumulate WIP in front of each machine in advance to avoid stock-outs. In addition, since only one robot was used, if multiple machines were available at the same time, they could not be supplied simultaneously. To address this challenge, they proposed a sequential part production matrix method.

Cycle time is often subject to uncertainty. If the distribution of production conditions can be specified, such uncertainty can be handled. Zhao et al. [13] scheduled random customer orders to minimize cycle times while taking into account setup times. In their experiment, the arrival times of customer orders, the processing times of workpieces, and the setup times of machines were all uncertain and followed specific distributions. They proposed three scheduling algorithms to minimize the expected average cycle time.

For some wafer fabrication processes, such as atomic layer deposition (ALD), wafers need to visit the same equipment multiple times. According to Qiao et al. [14], applying the existing scheduling method based on the swap-based strategy in such systems led to three-wafer cycle scheduling. To further improve the scheduling performance, Qiao et al. designed a timed Petri net to analyze the characteristics of three-wafer scheduling, and then developed a one-wafer schedule based on the new swap-based strategy, which further shortened cycle times.

5.4 Yield Improvement

Improving the yield can help reduce cycle time by avoiding rework [15]:

$$C_j = \mathrm{TP}_j + W_j + r(Y_j) \cdot R_j \tag{5.11}$$

where C_j, TP_j, W_j, Y_j and R_j indicate the cycle time, total processing time, waiting time, yield and repair time of job j, respectively. $r()$ determines whether job j needs to be reworked:

$$r(Y_j) = \begin{cases} 1 & \text{if} \quad Y_j < \xi \\ 0 & \text{otherwise} \end{cases} \tag{5.12}$$

ξ is the rework threshold. Therefore, according to Eqs. (5.11) and (5.12), C_j decreases as Y_j increases. However, this effect is limited. In lean manufacturing, cycle time is often matched to takt time [16] to prevent inventory buildup, which helps control cycle time [17]. In addition, if the same machine is used for performing rework on this job, the waiting times of subsequent jobs will be further lengthened:

$$C_j = \mathrm{TP}_j + W_j(\{(Y_l, R_l) | l = 1 \sim j - 1\}) + r(Y_j) \cdot R_j \tag{5.13}$$

Example 5.4 A simple flowshop consists of three consecutive machines, the details of which are presented in Table 5.7. Three jobs are released into the flowshop. Their production processes are summarized in Table 5.8. The average cycle time is 74.7 (mins). If the yield at Machine #2 is improved by 3%, the production processes are shown in Table 5.9. The average cycle time reduces to 73.3 (mins), an improvement of 1.7%.

For defective items, Sonntag and Kiesmüller [15] compared two different strategies: disposal or rework. Disposal did not cause any time delay, while the rework process required a positive rework time. Therefore, disposal did not extend the cycle time, while rework did. However, to compensate for the shortfall caused by the disposal, another job might need to be release into the factory, which delayed the makespan. Sonntag and Kiesmüller then built and solved an optimization model to support planners in deciding whether to dispose of defective items or to rework them.

Table 5.7 Details of jobs release to the flowshop

Machine #	Processing time (min)	Yield	Rework threshold	Rework time (min)
1	15	100%	–	–
2	20	$N(85\%, 13\%)^*$	78%	20% of processing time
3	17	100%	–	–

* Normal distribution

Table 5.8 Production processes of three jobs

Job #	Step #	Start time	Completion time	Yield (%)	Rework?	Rework time
1	1	0	15	100	No	0
1	2	15	35	73	Yes	4
2	1	20	35	100	No	0
2	2	35	55	80	No	0
1	3	39	56	100	No	0
3	1	40	55	100	No	0
3	2	55	75	77	Yes	4
2	3	55	72	200	No	0
3	3	79	96	300	No	0

Table 5.9 Production processes after yield improvement

Job #	Step #	Start time	Completion time	Yield (%)	Rework?	Rework time
1	1	0	15	100	No	0
1	2	15	35	76	Yes	4
2	1	20	35	100	No	0
2	2	35	55	83	No	0
1	3	39	56	100	No	0
3	1	40	55	100	No	0
3	2	55	75	80	No	0
2	3	55	72	200	No	0
3	3	75	92	300	No	0

Haji et al. [18] considered a single-machine system that produced some defective parts. Each time period consisted of multiple identical cycles. Defective parts produced in all cycles were accumulated and reworked in another (the last) cycle of this time period (called the rework cycle). To facilitate production and resource planning, they sought to make the rework cycle the same length as other cycles, which is similar to the concept of matching cycle time with required time in lean manufacturing, conducive to the control and reduction of cycle time.

Chiu et al. [19] proposed a simplified method to determine the cycle times of jobs in a multi-product finite production rate (FPR) system with random reworks. By eliminating the randomness of reworks, the cycle times could be significantly controlled. Subsequently, by reducing the frequency and duration of reworks, the cycle times were effectively shortened.

Leachman and Ding [2] discussed the impact of long cycle times on wafer die yield, where the wafer fabrication process or equipment deviated from specifications but was not detected in-line, resulting in wafer defects not being discovered until they were tested in the probing area after fabrication was completed. This problem became more serious due to longer cycle times because more wafers in fabrication were affected. They proposed an analytical formula to quantify the yield loss as a function of cycle time, deviation probability, and kill rate, which also provided a means to quantify the revenue gain from cycle time reduction.

5.5 Controlling Production Conditions

Cycle time is affected by many factory conditions, such as the WIP level, queue lengths, release policy and rate, job sequencing and scheduling method, time between machine failures, and repair time. Therefore, adjusting these production conditions can be considered a feasible way to shorten cycle time. However, not all production conditions can be easily changed, as illustrated in Fig. 5.8.

The impact of the release policy on cycle time is usually studied by selecting the optimal release policy from several alternatives (e.g., POISS, DETERMIN, CL, WR1(C), WR2(A, B), and WR3(C)) (see Table 5.10) to reduce cycle time [20]. From this perspective, Chen [4] established a systematic procedure to guide the formulation of management measures to change the production conditions of the factory, aiming to shorten the cycle time of jobs in the factory.

The cycle times of jobs can be shortened by eliminating unnecessary waiting time at each machine (depending on the number of jobs waiting for that machine). Most research attempted to achieve this by applying new scheduling methods. In contrast, another approach is to reduce the flow of jobs to highly congested machines by stopping jobs from their upstream machines, for which the objective function is to

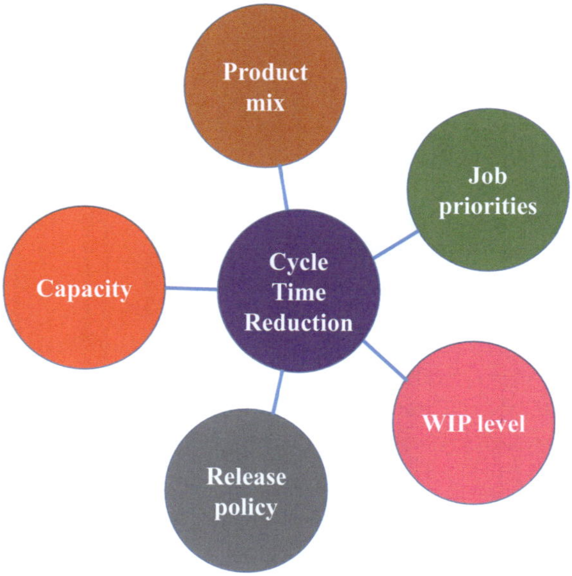

Fig. 5.8 Production conditions that can be changed for cycle time reduction

Table 5.10 Various release policies

Release policy	Full name	Meaning
POISS	Poisson release policy	Interarrival times follow an exponential distribution
DETERMIN	Deterministic release policy	Interarrival times are equal to the total time available divided by the number of jobs to be processed
CL	Close loop release policy	The number of jobs in the manufacturing system is held constant
WR	Work regulating release policy	Jobs are released into the manufacturing system whenever the number of jobs for the bottleneck falls below a pre-specified level

minimize the average cycle time of jobs by temporarily stopping inflows to highly congested machines:

$$\text{Min } Z = \frac{1}{n}\sum_{j=1}^{n} \Delta C_j \tag{5.14}$$

where ΔC_j is the cycle time reduction of job j, $j = 1, \ldots, n$. The effects of these measures are reflected in factory conditions such as fewer jobs on machines, lower work-in-process (WIP) levels, fewer jobs on processing routes, and fewer jobs waiting for bottleneck machines. These representative factory conditions are denoted by ψ_g, $g = 1, \ldots, G$. Since the actions are short term, their effects on the factory conditions are limited:

$$d\left(\psi_g(t), \psi_g(t_c)\right) \leq \nabla_{\psi_g} \tag{5.15}$$

where t is time, t_c is the current time, ∇_{ψ_g} is the maximal adjustment that can be made to ψ_g, and $d()$ is the distance function.

The systematic process established by Chen [4] consists of five steps (see Fig. 5.9):

Step 1. Identify controllable factors that affect cycle time, such as job attributes and factory conditions.
Step 2. Use machine learning (ML) or deep learning (DL) models to fit the relationship between controllable factors and cycle time.
Step 3. Plan actions to control factory conditions to shorten cycle time.
Step 4. Evaluate the effectiveness of the actions to shorten cycle time.
Step 5. If the reduction in the average cycle time is sufficient, stop; otherwise, return to Step 2.

Fitting the relationship between cycle time and factory conditions is not easy. In particular, the variability of cycle time is difficult to account for, which can lead to inadequate control of factory conditions. Figure 5.10 provides an example [4].

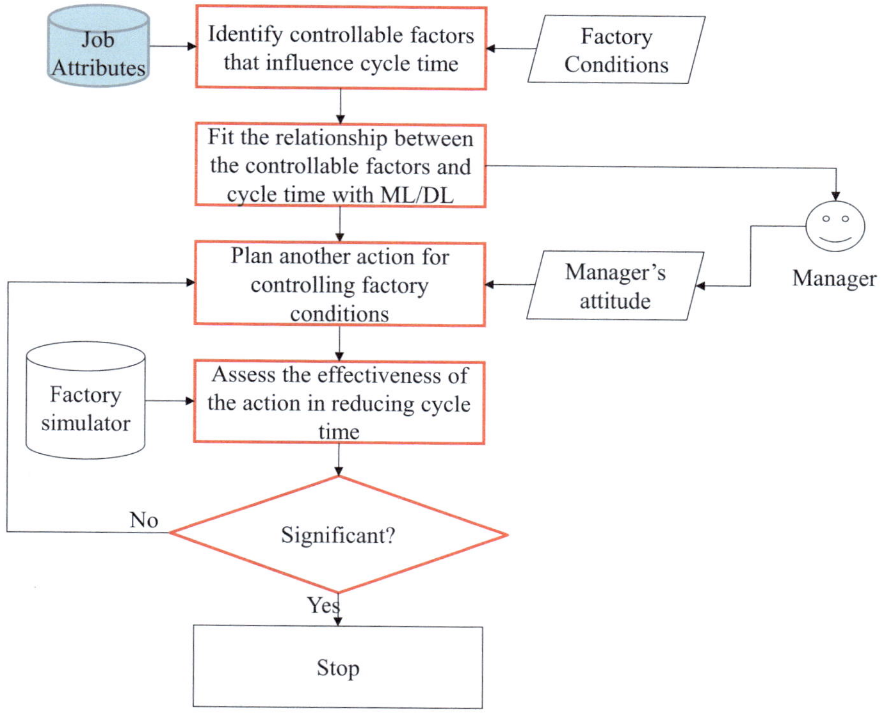

Fig. 5.9 Systematic procedure for controlling production conditions to reduce cycle time

According to the curve fitted by the ML application, when the factory condition change from d_1 to d_2, the cycle time is expected to decrease from c_1 to c_2. However, after the change, the cycle time increases from c_3 to c_4.

To address this problem, the lower and upper bounds of cycle time were considered by Chen [4] in planning actions to control factory conditions using a fuzzy ML application, as illustrated in Fig. 5.11. When factory conditions change from d_1 to

Fig. 5.10 Adjusting factory conditions to reduce cycle time

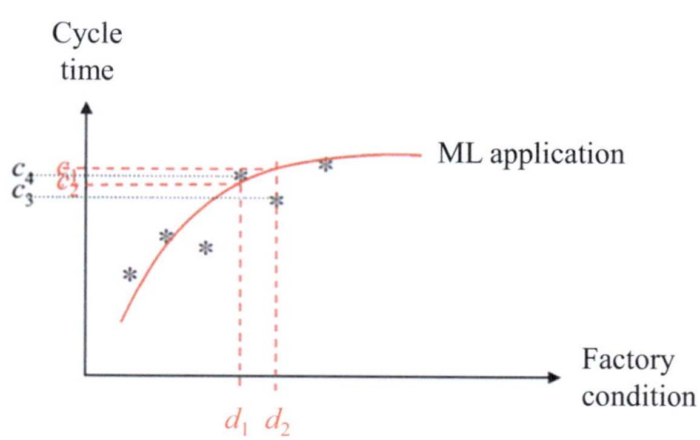

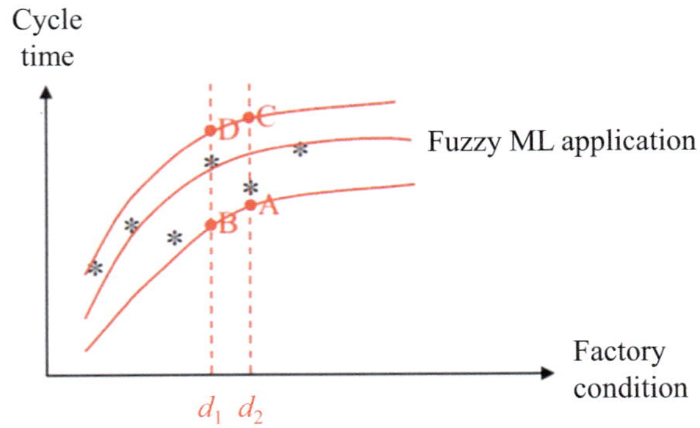

Fig. 5.11 Considering the lower and upper bounds of cycle time in planning actions to control factory conditions

d_2, the cycle time may change from A to B (or D), or from C to B (or D). Among these possibilities, only one is correct. Therefore, the fuzzy ML application estimates the cycle time of a job with a triangular fuzzy number (TFN). The three corners of the TFN represent the lower bound, center (most likely value), and upper bound of the cycle time, respectively. These values are then presented to the manager. An optimistic manager expects a change from C to B, whereas a pessimistic manager expects a change from A to D. The manager then plans cycle time reduction actions based on his or her expectation.

According to Little's law, a reduction in WIP will lead to a reduction in cycle time while maintaining the same output level. From this perspective, predicting and controlling the WIP level is crucial [21]. Accurately predicting the future WIP level is also an important task for factory control [22]. However, there is little research on predicting the WIP level of tool sets with tool dedication and waiting time constraints. In addition, the existing queuing models for predicting the WIP level of tool sets in a wafer fab have limitations in practical applications. To this end, Kuo et al. [21] constructed an ANN to predict the WIP level of tool sets based on historical production data and equipment data, thereby controlling the corresponding WIP levels of different tool sets and shortening cycle times without affecting the output.

For a similar purpose, Chen [22] proposed a fuzzy linear regression (FLR)-partial consensus fuzzy intersection (PCFI)-radial basis function network (RBF) method to predict the WIP level in a wafer fab. In the FLR-PCFI-RBF method, they formed a virtual expert committee instead of convening many experts in the field. For each virtual expert, a corresponding FLR equation was used to predict the WIP level. Each FLR equation was fitted by solving two equivalent nonlinear programming (NLP) problems based on the opinion of the virtual expert. To aggregate these fuzzy WIP level predictions, a two-step aggregation mechanism was adopted. First, PCFI was applied to aggregate the fuzzy predictions into a polygonal fuzzy number to improve the prediction precision. Then, an RBF was constructed to defuzzify the polygonal fuzzy number and generate a representative/crisp value to improve the prediction accuracy.

When jobs in a factory have different priorities, job priority becomes another production condition that can be adjusted to reduce cycle time. Raising the priority of a job will obviously speed it up, thus reducing its cycle time, while other jobs in the factory may be delayed. Therefore, the adjustment of job priority is rather a cycle time control measure than a cycle time reduction measure.

The percentage of hot jobs in the product mix of a factory also affects the cycle times of all jobs in the factory. Gupta et al. [23] aimed to minimize the average cycle time of hot jobs in batch processes. They proposed three scheduling strategies based on look-ahead batching (LAB), where scheduling decisions were made by considering the arrival times of incoming batches, which was easy to predict in a computer-integrated manufacturing environment. Simulation results showed that the LAB-based scheduling strategies reduced the average cycle time of hot batches by 10% without affecting the cycle time variation. In addition, the cycle time performance of jobs with other priorities was not affected when the fab utilization was low.

Increasing production capacity by purchasing more machines is obviously the most direct way to shorten cycle time. If the production capacity of a certain step is doubled, then the cycle time of that step will be halved, and cumulatively the cycle time will be shortened.

5.6 Three-Dimensional Printing

Three-dimensional (3D) printing is another way to shorten cycle times. With 3D printing, products are manufactured in a single step, rather than passing through multiple different machines. No transportation between machines or setup of each machine is required, which helps reduce cycle times more than traditional methods. Additionally, in a 3D printing-based manufacturing system, different parts share the same raw materials, also eliminating machine changeovers and further shortening cycle times.

The 3D computer-aided design (CAD) files of products can be easily transmitted and exchanged via the Internet, and products can be manufactured without recipes, eliminating the need for exchanging recipes. Therefore, several 3D printing facilities scattered in a region can form a ubiquitous manufacturing (UM) system. Customers in different locations use the same Web interface to place orders, and then the UM system automatically distributes these orders to 3D printing facilities to complete all orders as soon as possible or in time.

In the view of Chen [24], human-induced uncertainty and premature termination are two complex factors that hinder the effectiveness of UM systems based on 3D printing. They addressed these issues by defining fuzzy parameters and deriving slacks to determine whether to restart a prematurely terminated 3D printing process in the same 3D printing facility. According to the experimental results, their methodology reduced cycle time by 9% compared to two existing methods that distributed more parts to closer or faster 3D printing facilities.

5.7 Utilizing Cloud Manufacturing Capacity

Long cycle time is caused by too much waiting time, which is due to insufficient short-term capacity. Factories can use cloud capacity to solve this problem [25] (see Fig. 5.12):

Step 1. Schedule jobs in the factory.
Step 2. Identify jobs with relatively long cycle times, for example, jobs with cycle times much longer than their total processing times.
Step 3. For each identified job, find its operation that, if forwarded, can significantly reduce its cycle time.
Step 4. Plan the processing of the operation using cloud capacity.

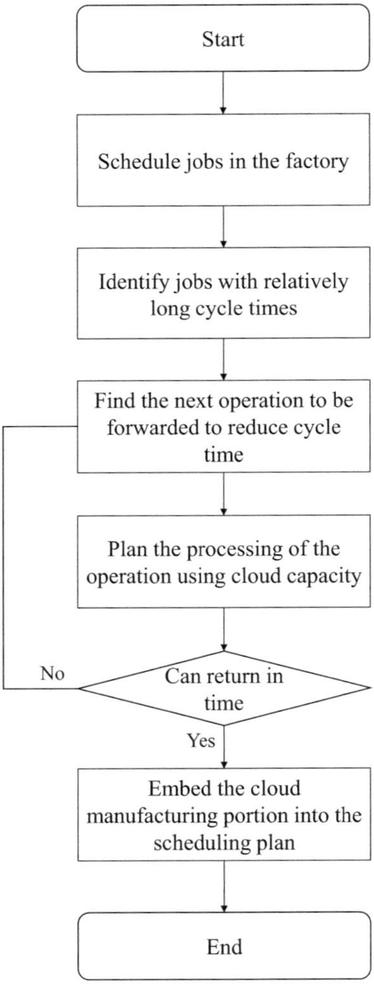

Fig. 5.12 Procedure for using cloud capacity to reduce the cycle time of a job

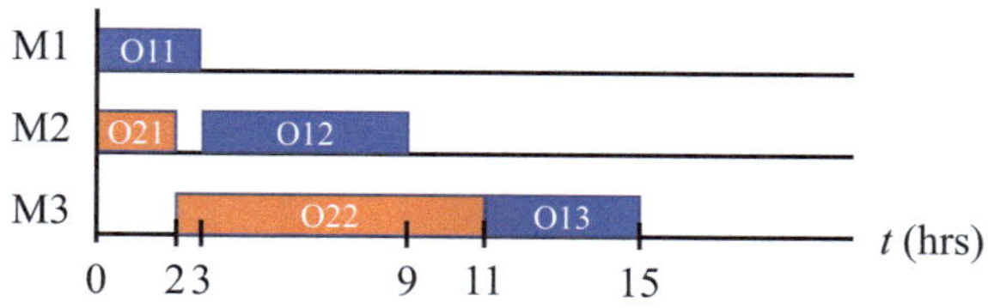

Fig. 5.13 Scheduling results of a manufacturing system

Step 5. If the job can be returned in time, embed the cloud manufacturing portion into the scheduling plan; otherwise, return to Step 3.

Example 5.5 As shown in Fig. 5.13, a factory has scheduled its two jobs. The cycle time of Job #2 is 11, which cannot be shortened. In contrast, the cycle time of Job #1 can be improved because the job waited for 2 h for the next machine in the manufacturing process. To solve this problem, cloud capacity can be resorted, that is, a machine that can process O13 can be found in another factory. The condition is that all actions of shipping Job #1 to the external factory, processing O13, and shipping it back to the original factory should be completed before the 15th hour (see Fig. 5.14). In this way, the cycle time of Job # can be shortened. The cloud manufacturing plan is presented in Table 5.11.

Ren et al. [26] defined cloud manufacturing as a smart networked manufacturing model based on cloud computing, aiming at meeting the growing demand for product personalization, broader global cooperation, knowledge-intensive innovation, and increased response agility, including cycle time reduction. In cloud manufacturing,

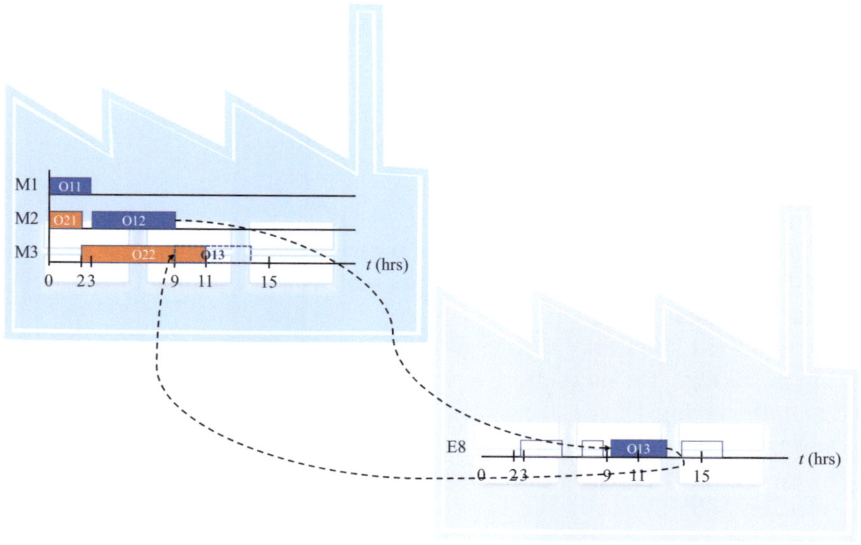

Fig. 5.14 Utilizing cloud capacity to reduce cycle time

Table 5.11 Cloud manufacturing plan

Time (hr)	Action
9 ~ 9.2	Job #1 is transported to another factory
9.2 ~ 12.5	O13 is processed using cloud manufacturing capacity
12.5 ~ 13.3	Job #1 is transported back to the original factory
13.3	Job #1 is completed. Its cycle time has been reduced by 1.7 h without affecting Job #2

customers can access a shared pool through the network and enjoy on-demand services that support the entire life cycle of a product. In the shared pool, distributed manufacturing resources are virtualized and uniformly managed in a configurable and optimized manner.

Wang et al. [27] classified existing cloud manufacturing applications into two categories: applications of cloud computing technologies to manufacturers and cloud-based manufacturing systems. The case in Example 5.5 belongs to the first category.

Zhang et al. [28] solved the planning and scheduling problem of a manufacturing system consisting of parallel batch machines located at different locations and with different capabilities. It was assumed that jobs had arbitrary release times, different job sizes, and different processing times. The goal was to minimize the cycle time equal to the sum of production completion time and delivery time, and a mixed integer linear programming (MILP) model was optimized for this purpose. Since this problem was NP-hard, an effective heuristic algorithm based on particle swarm optimization (PSO) was proposed to solve it.

According to Chen [29], the benefits of cloud manufacturing included cost savings, efficiency (such as shorter cycle times), additional data analysis capabilities, advanced automation (such as cloud robotics), flexibility, and closer partner relationships.

5.8 Pull Manufacturing

There are several techniques in **lean manufacturing** (or **lean enterprise management**) that can be used to reduce cycle time, such as **pull manufacturing**, **VSM**, and **lean six sigma** [30].

According to Little's law [31], the average cycle time is proportional to the WIP level in the factory. The most effective way to control the WIP level in a factory is **pull manufacturing**, that is, an upstream machine can complete the task only when the downstream machine is available and then send it to the downstream machine. This way, the queue in front of each machine is empty or very short, accumulating to a very low WIP.

Example 5.6 Five jobs are released to a manufacturing system consisting of two machines for processing. Their processing times are summarized in Table 5.12.

Figure 5.15 compares the cycle times of these jobs when the manufacturing system adopts the "pull" and "push" manufacturing mechanisms.

Ghrayeb et al. [32] viewed a make-to-order assembly plant as a **hybrid push–pull system**, in which raw materials can be converted into common semi-finished products at some point in the production process, and the next downstream process is triggered by customer orders. Production at earlier upstream machines was controlled by push production, while production at later downstream machines was controlled by pull production. The push part of the hybrid system suffered from higher inventory and longer cycle time, but achieved higher output rate. In contrast, the pull part had lower inventory and shorter cycle time, but suffered from lower output rate. They tried to determine the optimal split point to minimize the sum of inventory holding costs and delivery costs.

Wang et al. [33] solved the multi-site production planning problem of a semiconductor wafer ingot manufacturing company consisting of five factories distributed in different countries. A two-stage planning model based on a **pull manufacturing** strategy was proposed. In the first stage, every available machine sent a request to the pull logic controller to select suitable orders from the central queue and accumulate them into the production kanban. In the second stage, the completed orders were assigned to customers so that the total transportation cost was minimized. A linear programming (LP) problem was solved for this purpose.

Table 5.12 Processing times of five jobs

Job #	Processing time on machine #1 (min.)	Processing time on machine #2 (min.)
1	8	14
2	9	17
3	6	11
4	9	17
5	5	8

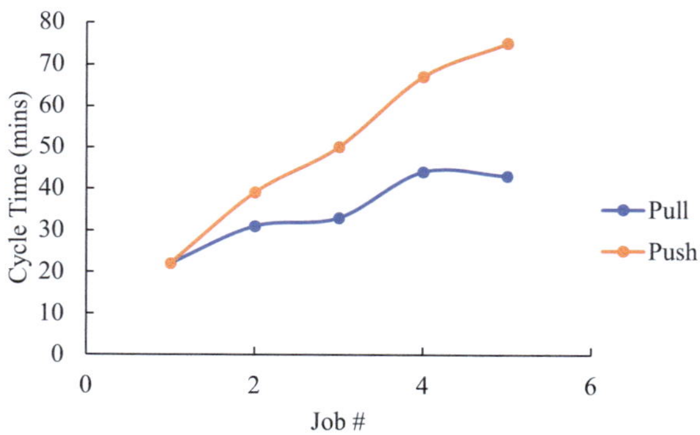

Fig. 5.15 Comparison of cycle times when pull and push manufacturing mechanisms are employed

5.9 VSM

VSM is another lean manufacturing technique used to reduce cycle time (see Fig. 5.16). VSM aims to record, analyze, and improve the information and material flow in a manufacturing process [34], thereby eliminating various wastes. VSM basically consists of six components: customers, production planning and control, suppliers, receiving, production steps, and delivery. Methods for improving VSM include pull production, adjusting storage space capacity, improving the operations of each workstation to shorten the processing time, and splitting or merging operations to reduce the cycle time of the manufacturing process. Unlike pull production that treats WIP, VSM reduces cycle time by regulating or matching the cycle times of all steps to the takt time.

Seth et al. [35] applied VSM to reduce the cycle time in a heavy industrial power transformer plant that featured an engineer-to-order (ETO) and high-mix low-volume (HMLV) production environment. The improvement team used **gemba (i.e., on-site) walks** and systematic questioning techniques to collect relevant data. Then, a **cause-and-effect diagram** was drawn to identify the key factors that led to poor product quality and long cycle times. Subsequently, **Taguchi method** was applied to optimize the values of these key factors to minimize cycle time or maximize product quality.

In the study by Tyagi et al. [36], VSM was used to shorten the cycle time of a product development process that could last for several years. They used the development process of a gas turbine product as an example to illustrate and justify the VSM application. First, the analyst drew the current (as-is) VSM after gemba walks by subject matter experts. Then, they brainstormed to identify the root causes of wastes found during gemba walks. After formulating measures to eliminate all wastes/inefficiencies, the future (to-be) VSM was drawn. The application of VSM helped the product development team shorten the cycle time of the product development process by 50% and brought many intangible benefits.

Venkataraman et al. [37] applied VSM to reduce the cycle time of a crankshaft machining process. After collecting all the information about the current state of the machining process, the as-is VSM diagram was drawn. Subsequently, three types of improvements were implemented: eliminating and merging processes, optimizing process parameters, and improving tools and introducing CNC lathes. As a result, the cycle time of the crankshaft machining process was reduced by 40%. In addition, the number of product defects was also reduced, while the process capability was improved and the response speed to customer needs was accelerated.

Siva et al. [38] studied a case study of an airline company that attempted to shorten the cycle times of all maintenance projects. They first broke down maintenance services into four different levels. The first three levels were completed at the maintenance plant, while the fourth level, replacing parts in the aircraft engine, was completed at the customer's location with authorization from the customer. Initially, the average cycle time for each maintenance project was 61.5 h, which was successfully shortened after applying lean techniques such as VSM, takt time, poka-yoke, Jidoka, 5S, and Muda.

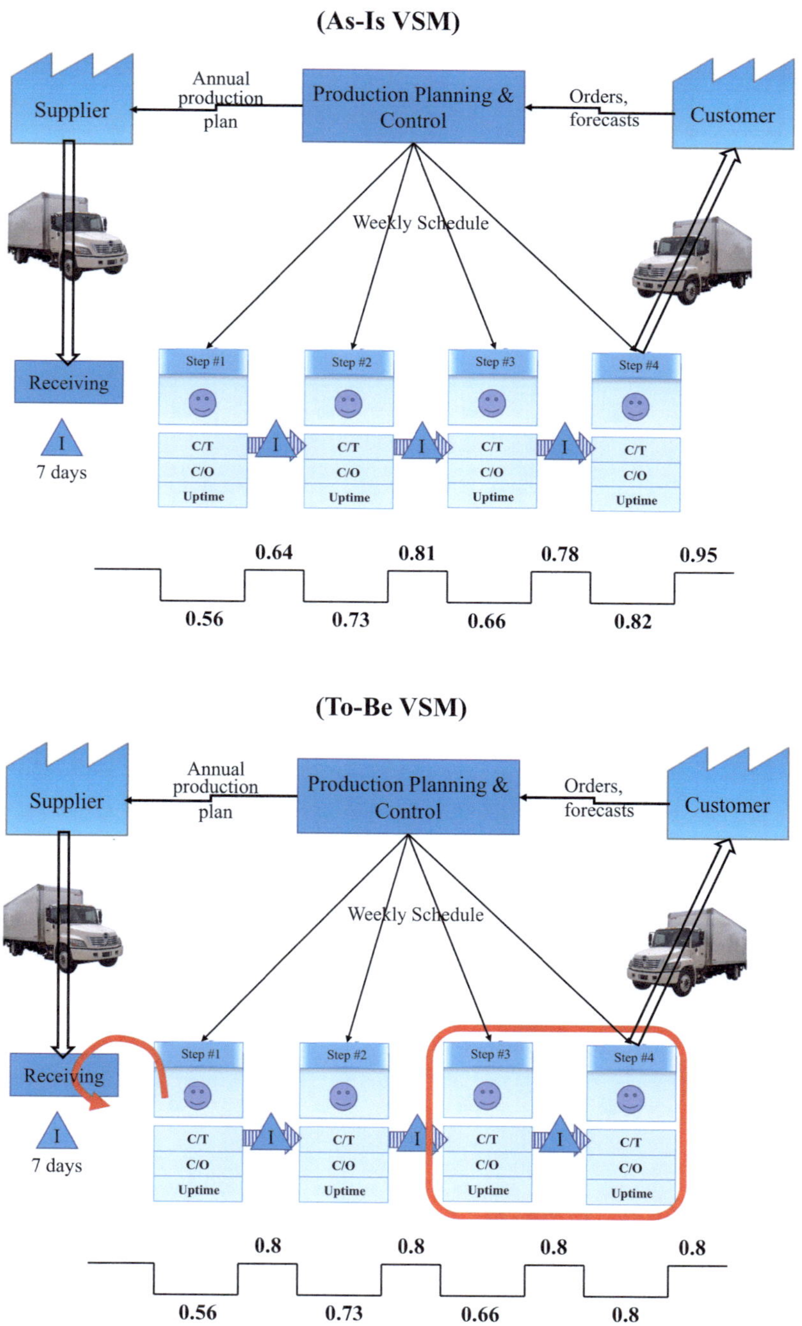

Fig. 5.16 VSM for cycle time reduction

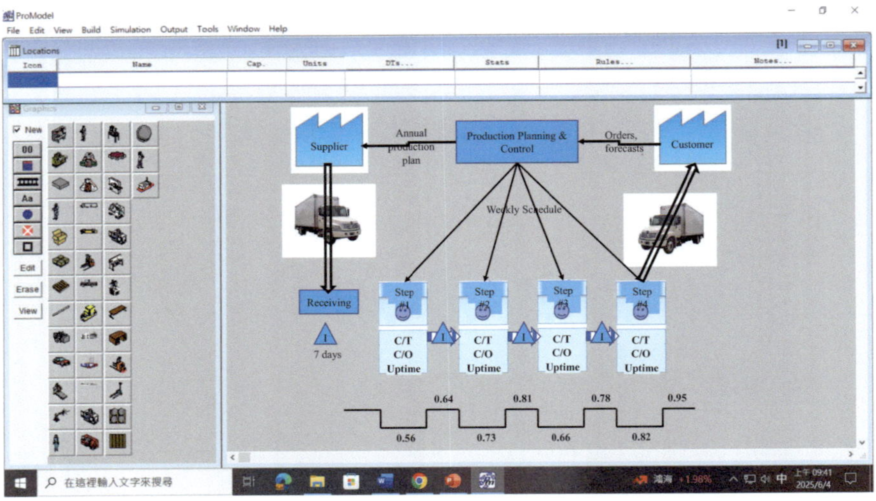

Fig. 5.17 VSM using simulation

VSM is not actually a tool for cycle time reduction, but rather a visualization tool that can show problems with a manufacturing system, including unnecessary inventory and waiting. However, in VSM, system diagrams of the current state and the future state need to be drawn, which is a time-consuming task. How to transition from the former to the latter is also non-systematic, mostly based on subjective experience and knowledge. To overcome this difficulty, Rasi et al. [39] mentioned the possibility of automatically generating VSM diagrams through simulation, where the components of the VSM diagram were placed on the layout of the manufacturing system. Improvement concepts such as error-proofing, continuous flow production, pull systems, and supermarkets with Kanban featured different scenarios for which simulations were run. The results were then displayed on the system layout, including the required information (cycle times, processing times, inventory levels, etc. after improvement) displayed on a VSM diagram (see Fig. 5.17).

5.10 Lean Six Sigma

Lean six sigma (LSS) is a management approach that combines lean management and six sigma, aiming to improve an organization's operations by reducing waste, lowering variation, improving efficiency, and adding to customer value. Lean manufacturing improves the planning, control, and management of a manufacturing system by using simple and effective tools such as kanbans, pacemaker, value stream mapping, 5 s, just-in-time (JIT), standard operating procedures, load leveling, pull manufacturing, and others [40]. Six sigma is a process that improves product quality by identifying and eliminating processes that cause defects and reduces the degree of variation in production and business processes [41]. The concept of six sigma originates from statistics that most quality measures follow normal distribution under which 99.99966% of the products produced (3.4 defects per million) are free of quality problems, where the quality measure is allowed to deviate from the nominal (target) value by 1.5 standard deviations, therefore the range is six standard deviations. Six sigma goals are not only established for production processes, but also extended to management and engineering processes.

Define, measure, analyze, improve, and control (DMAIC) is a five-step methodology commonly used in six sigma to guide improvement projects:

- Define: Identify the problem, objectives, and scope.
- Measure: Collect data and quantify the problem.
- Analyze: Analyze the data to find the root cause.
- Improve: Develop and implement solutions.
- Control: Monitor and maintain improvements.

Ismail et al. [20] applied DMAIC to identify waste and reduce cycle time in biopharmaceutical production. They broke down the process into detailed steps and then classified the steps as value-added or non-value-added activities. In contrast, in traditional motion study, activities are usually broken down into motions of five categories: storage, transportation, delay, operation, and inspection. Among them, only operation and inspection are value-added motions (see Fig. 5.18).

Step	Operation	Transport	Delay	Inspection	Storage
Step #1	●	⇨	D	□	▽
Step #2	○	⇨	D	□	▼
Step #3	●	⇨	D	□	▽
…					

(Motion study)

Step	Value-added	Non-value-added
Step #1	√	
Step #2		×
Step #3	√	
…		

(Lean manufacturing)

Fig. 5.18 Non-value-added operations in motion study and lean manufacturing

In Ismail et al. [20], the sum of the cycle times for all non-value-added activities was the greatest benefit in reducing cycle time. Based on the experimental results, about 54% of the steps were considered waste or non-value-added activities and could be easily eliminated. To achieve better results and continuous improvement, some activities were introduced, such as special carts or trolleys for product transportation, closing all exposed pipes or tubes with special headers, installing a heat exchanger to increase cooling capacity, modifying the room to increase the working area and properly arranging the equipment.

5.11 Industry 4.0 Applications in Cycle Time Reduction

Factories around the world are committed to shortening cycle times. To achieve this goal, it is necessary to fit the relationship between the attributes of the job (or production conditions) and the cycle time. Based on the fitted relationship, these job attributes (or production conditions) can be adjusted if possible to shorten the cycle time of the job. However, many possible production conditions can affect the cycle time of a job. In addition, a large and complex production system contains thousands or even tens of thousands of jobs, which makes the prediction and shortening of the cycle time a complex big data problem. To solve this problem, **big data analytics** in Industry 4.0 is a useful tool. In addition, **edge computing** is expected to facilitate the downsizing of big data by processing raw data at the forefront of the information flow [40, 41].

Furthermore, as described in Sect. 5.7, in the short term, when self-owned capacity in a factory is insufficient, cloud capacity can be used to avoid excessive waiting times and reduce cycle times. To do this, a prerequisite is to know the availability of machines in other factories as quickly as possible. Otherwise, the machines available in a factory may be quickly occupied by jobs in this factory or from other factories. For this purpose, **Internet-of-machines (IoM)** applications are very useful, which can connect machines in different factories and inform each other about their availability (see Fig. 5.19).

In addition, if a machine suddenly fails, the jobs waiting for that machine will be delayed, thereby extending the cycle time of these jobs. These jobs also need to be assigned to other machines, which will cause the jobs waiting for these machines to be delayed as well. In other words, avoiding such events, or at least mitigating the impact of such events, helps to shorten the cycle times of the jobs. To this end, **predictive maintenance (PdM)** has emerged. PdM is the use of sensors to monitor the condition of a machine (such as temperature, pressure, and noise level) to estimate the time when the machine will next fail [42]. PdM is an application of cyber-physical systems (CPSs) in Industry 4.0. If the estimated time is earlier than expected, a periodic maintenance of the machine can be scheduled in advance.

Example 5.7 Four jobs are to be processed on two machines, M1 and M2. The data of these jobs are presented in Table 5.13. Without unexpected machine failures, the

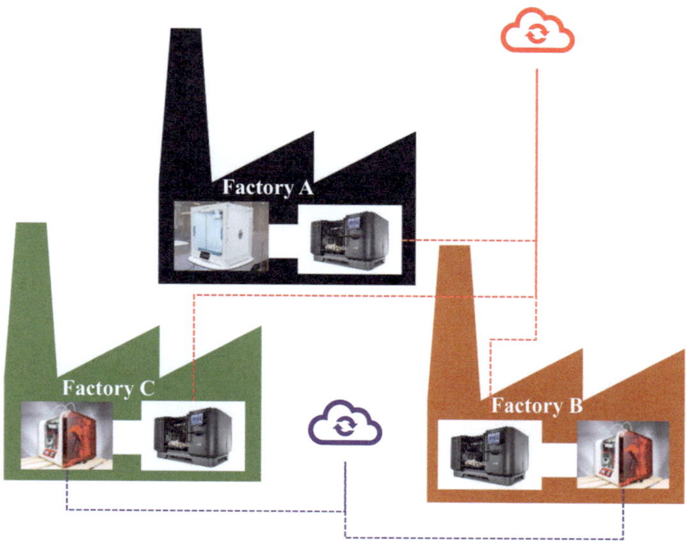

Fig. 5.19 IoM application

Table 5.13 Data of jobs

Job #	Release time	Estimated completion time	Estimated cycle time
J3	3	26	23
J4	5	23	18
J5	6	27	21
J6	7	30	23

average cycle time of the four jobs is 21.25 (see Fig. 5.20). M1 is predicted to fail at time 18 after analyzing the sensing results, which happens during the processing of Job #3, as shown in Fig. 5.21. If a PdM is not arranged for M1, after M1 failures, J3 and J6 need to be reassigned, as shown in Table 5.14. The average cycle time becomes 24.5, as shown in Fig. 5.22. To avoid that, a periodic maintenance is arranged for M1. As a result, the average cycle time changes to 23.75, 3% less than that without PdM, as shown in Table 5.15.

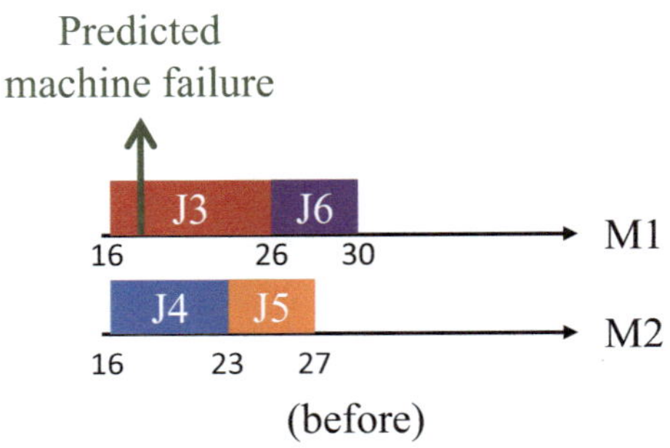

Fig. 5.20 Before rescheduling

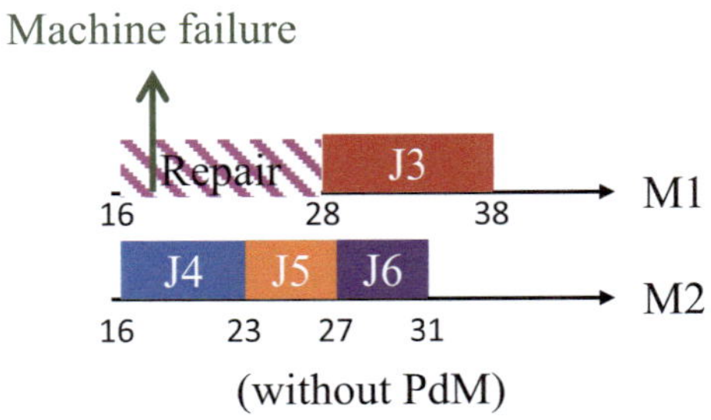

Fig. 5.21 Without PdM

Table 5.14 Scheduling results without PdM

Job #	Release time	Estimated completion time	Estimated cycle time
J3	3	38	35
J4	5	23	18
J5	6	27	21
J6	7	31	24

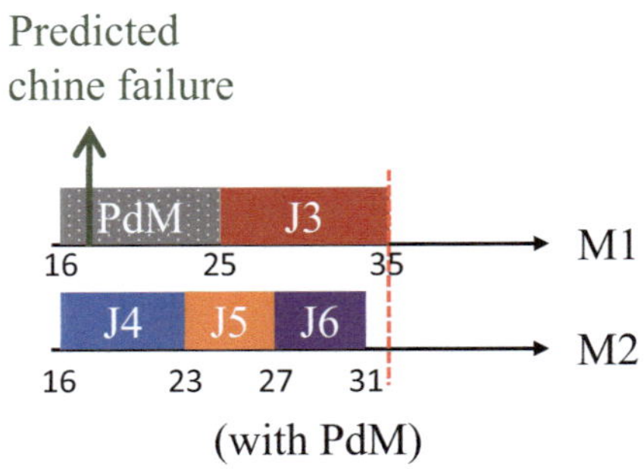

Fig. 5.22 With PdM

Table 5.15 Scheduling results with PdM

Job #	Release time	Estimated completion time	Estimated cycle time
J3	3	35	32
J4	5	23	18
J5	6	27	21
J6	7	31	24

References

1. M. Attaran, The rise of 3-D printing: The advantages of additive manufacturing over traditional manufacturing. Bus. Horiz. **60**(5), 677–688 (2017)
2. R.C. Leachman, S. Ding, Excursion yield loss and cycle time reduction in semiconductor manufacturing. IEEE Trans. Autom. Sci. Eng. **8**(1), 112–117 (2010)
3. I. Taifa, T. Vhora, Cycle time reduction for productivity improvement in the manufacturing industry. J. Ind. Eng. Manage. Stud. **6**(2), 147–164 (2019)
4. T. Chen, A fuzzy back-propagation network approach for planning actions to shorten the cycle time of a job in dynamic random access memory manufacturing. Neural Comput. Appl. **26**, 1813–1825 (2015)
5. T. Chen, Y.-C. Wang, H.-R. Tsai, Lot cycle time prediction in a ramping-up semiconductor manufacturing factory with a SOM-FBPN-ensemble approach with multiple buckets and partial normalization. Int. J. Adv. Manuf. Technol. **42**(11–12), 1206–1216 (2009)
6. T. Chen, Y.-C. Lin, A fuzzy-neural fluctuation smoothing rule for scheduling jobs with various priorities a semiconductor manufacturing factory. Internat. J. Uncertain. Fuzziness Knowl.-Based Syst. **17**(3), 397–417 (2009)
7. M.L. Pinedo, *Scheduling: Theory, Algorithms, and Systems* (Springer, 2022)
8. Y.T. Kao, C.M. Chang, S.C. Chang, Do we still need daily production target setting in fully automated fabs?, in *2014 e-Manufacturing & Design Collaboration Symposium* (2014), pp. 1–4
9. T. Chen, An effective dispatching rule for bi-objective job scheduling in a wafer fabrication factory—considering the average cycle time and the maximum lateness. Int. J. Adv. Manuf. Technol. **67**(5–8), 1281–1295 (2013)
10. T.C.T. Chen, Job sequencing and scheduling, in *Production Planning and Control in Semiconductor Manufacturing: Big Data Analytics and Industry 4.0 Applications* (2022), pp. 41–62
11. T. Chen, Optimized fuzzy-neuro system for scheduling wafer fabrication. J. Sci. Ind. Res. **68**, 680–685 (2009)
12. S. Mosallaeipour, M.G. Nejad, S.M. Shavarani, R. Nazerian, Mobile robot scheduling for cycle time optimization in flow-shop cells, a case study. Prod. Eng. Res. Devel. **12**(1), 83–94 (2018)
13. Y. Zhao, X. Xu, H. Li, Y. Liu, Stochastic customer order scheduling with setup times to minimize expected cycle time. Int. J. Prod. Res. **56**(7), 2684–2706 (2018)
14. Y. Qiao, N. Wu, M. Zhou, A Petri net-based novel scheduling approach and its cycle time analysis for dual-arm cluster tools with wafer revisiting. IEEE Trans. Semicond. Manuf. **26**(1), 100–110 (2012)
15. D. Sonntag, G.P. Kiesmüller, Disposal versus rework–inventory control in a production system with random yield. Eur. J. Oper. Res. **267**(1), 138–149 (2018)
16. M.A. Millstein, J.S. Martinich, Takt time grouping: implementing kanban-flow manufacturing in an unbalanced, high variation cycle-time process with moving constraints. Int. J. Prod. Res. **52**(23), 6863–6877 (2014)
17. R.M. Ali, A.M. Deif, Dynamic lean assessment for takt time implementation. Procedia CIRP **17**, 577–581 (2014)

18. B. Haji, A. Haji, A. Rahmati Tavakol, Scheduling accumulated rework in a normal cycle: optimal batch production with minimum rework cycles. J. Ind. Syst. Eng. **2**(3), 236–249 (2008)
19. Y.S.P. Chiu, H.D. Lin, C.T. Tseng, S.W. Chiu, Determining cycle time for a multi-product FPR model with rework and an improved delivery policy by alternative approach. Int. J. Eng. Model. **29**(1–4), 27–36 (2016)
20. J. Teixeira, Release early, release often and release on time. An empirical case study of release management, in *IFIP International Conference on Open Source Systems* (2017), pp. 167–181
21. C.J. Kuo, C.F. Chien, J.D. Chen, Manufacturing intelligence to exploit the value of production and tool data to reduce cycle time. IEEE Trans. Autom. Sci. Eng. **8**(1), 103–111 (2010)
22. T. Chen, The FLR-PCFI-RBF approach for accurate and precise WIP level forecasting. Int. J. Adv. Manuf. Technol. **63**, 1217–1226 (2012)
23. A.K. Gupta, V.K. Ganesan, A.I. Sivakumar, Hot lot management: minimizing cycle time in batch processes, in *2004 IEEE International Engineering Management Conference*, vol. 3 (2004), pp. 1217–1221
24. T.C.T. Chen, Fuzzy approach for production planning by using a three-dimensional printing-based ubiquitous manufacturing system. AI EDAM **33**(4), 458–468 (2019)
25. T.C.T. Chen, Application of ubiquitous manufacturing to a conventional manufacturer, in *3D Printing and Ubiquitous Manufacturing* (2020), pp. 13–27
26. L. Ren, L. Zhang, L. Wang, F. Tao, X. Chai, Cloud manufacturing: key characteristics and applications. Int. J. Comput. Integr. Manuf. **30**(6), 501–515 (2017)
27. X.V. Wang, L. Wang, A. Mohammed, M. Givehchi, Ubiquitous manufacturing system based on cloud: a robotics application. Rob. Comput.-Integr. Manuf. **45**, 116–125 (2017)
28. H. Zhang, K. Li, C. Chu, Z.H. Jia, Parallel batch processing machines scheduling in cloud manufacturing for minimizing total service completion time. Comput. Oper. Res. **146**, 105899 (2022)
29. T.C.T. Chen, Introduction, in *3D Printing and Ubiquitous Manufacturing* (2020), pp. 1–12
30. A. Ismail, J.A. Ghani, M.N. Ab Rahman, B. Md Deros, C.H. Che Haron, Application of lean six sigma tools for cycle time reduction in manufacturing: case study in biopharmaceutical industry. Arab. J. Sci. Eng. **39**, 1449–1463 (2014)
31. S.H. Kim, W. Whitt, Statistical analysis with little's law. Oper. Res. **61**(4), 1030–1045 (2013)
32. O. Ghrayeb, N. Phojanamongkolkij, B.A. Tan, A hybrid push/pull system in assemble-to-order manufacturing environment. J. Intell. Manuf. **20**, 379–387 (2009)
33. P.S. Wang, T. Yang, L.C. Yu, Lean-pull strategy for order scheduling problem in a multi-site semiconductor crystal ingot-pulling manufacturing company. Comput. Ind. Eng. **125**, 545–562 (2018)
34. W. Shou, J. Wang, P. Wu, X. Wang, H.Y. Chong, A cross-sector review on the use of value stream mapping. Int. J. Prod. Res. **55**(13), 3906–3928 (2017)
35. D. Seth, N. Seth, P. Dhariwal, Application of value stream mapping (VSM) for lean and cycle time reduction in complex production environments: a case study. Prod. Plann. Control **28**(5), 398–419 (2017)
36. S. Tyagi, A. Choudhary, X. Cai, K. Yang, Value stream mapping to reduce the lead-time of a product development process. Int. J. Prod. Econ. **160**, 202–212 (2015)
37. K. Venkataraman, B.V. Ramnath, V.M. Kumar, C. Elanchezhian, Application of value stream mapping for reduction of cycle time in a machining process. Procedia Mater. Sci. **6**, 1187–1196 (2014)
38. R. Siva, M. Purusothaman, Y. Jegathish, Process improvement by cycle time reduction through lean methodology. IOP Conf. Ser.: Mater. Sci. Eng. **197**(1), 012064
39. R.Z.R. Rasi, R. Abdullah, N. Omar, S. Mohamed, Value stream mapping using simulation at metal manufacturing industry, in *International Conference on Industrial Engineering and Operations Management* (2014), pp. 2455–2464
40. T.C.T. Chen, Y.C. Wang, Basics in lean management, in *Artificial Intelligence and Lean Manufacturing* (2022), pp. 1–12
41. D.C. Montgomery, W.H. Woodall, An overview of six sigma, in *International Statistical Review/Revue Internationale de Statistique* (2008), pp. 329–346
42. Z. Tao, Q. Xia, Z. Hao, C. Li, L. Ma, S. Yi, Q. Li, A survey of virtual machine management in edge computing. Proc. IEEE **107**(8), 1482–1499 (2019)

MIX
Papier aus verantwortungsvollen Quellen
Paper from responsible sources
FSC® C105338
www.fsc.org

If you have any concerns about our products,
you can contact us on
ProductSafety@springernature.com

In case Publisher is established outside the EU,
the EU authorized representative is:
Springer Nature Customer Service Center GmbH
Europaplatz 3, 69115 Heidelberg, Germany

Printed by Libri Plureos GmbH
in Hamburg, Germany